Introduction to Mathematical Techniques Used in GIS

Introduction to Mathematical Techniques Used in GIS

Peter Dale

CRC PRESS

Boca Raton   London   New York   Washington, D.C.

Library of Congress Cataloging-in-Publication Data

Dale, Peter.
  Introduction to mathematical techniques used in GIS / by Peter Dale.
    p. cm.
  Includes bibliographical references and index.
  ISBN 0-415-33414-4 (alk. paper)
    1. Geography—Mathematics. 2. Geographic information systems. I. Title.

G70.23.D35 2004
910'.01'51--dc22   2004051924

This book contains information obtained from authentic and highly regarded sources. Reprinted material is quoted with permission, and sources are indicated. A wide variety of references are listed. Reasonable efforts have been made to publish reliable data and information, but the author and the publisher cannot assume responsibility for the validity of all materials or for the consequences of their use.

Neither this book nor any part may be reproduced or transmitted in any form or by any means, electronic or mechanical, including photocopying, microfilming, and recording, or by any information storage or retrieval system, without prior permission in writing from the publisher.

The consent of CRC Press does not extend to copying for general distribution, for promotion, for creating new works, or for resale. Specific permission must be obtained in writing from CRC Press for such copying.

Direct all inquiries to CRC Press, 2000 N.W. Corporate Blvd., Boca Raton, Florida 33431.

Trademark Notice: Product or corporate names may be trademarks or registered trademarks, and are used only for identification and explanation, without intent to infringe.

Visit the CRC Press Web site at www.crcpress.com

© 2005 by CRC Press

No claim to original U.S. Government works
International Standard Book Number 0-415-33414-4
Library of Congress Card Number 2004051924
Printed in the United States of America 1 2 3 4 5 6 7 8 9 0
Printed on acid-free paper

Preface

This book has been written for nonmathematicians who wish to understand some of the assumptions that underlie the manipulation and display of geographic information. It assumes very little basic knowledge of mathematics, but moves rapidly through a wide range of data transformations, outlining the techniques involved. Many of these are precise, building logically from certain underlying assumptions; others are based on statistical analysis and the pursuit of the optimum rather than the perfect and definite solution.

Mathematics has its own form of shorthand that often gets in the way of understanding what is going on. For those who are unfamiliar with mathematical notations, this can be daunting; but it cannot be avoided. In many cases, it can be kept to a minimum and in what follows, the derivations of some of the formulae are placed in boxes that can be digested at leisure without interrupting the narrative. But at the end of the day, compromise has had to be made and as the text progresses, there is an increasing use of symbols.

This spirit of compromise is most apparent in the selection of topics discussed. Many things have had to be left out — indeed, every chapter could be expanded to a full book and most would require several volumes in order to do justice to their subject. *Introduction to Mathematical Techniques used in GIS* is therefore a book that allows the reader to get started and then to turn to the many more informative texts that are available.

The text begins with an introduction to geographic data, but soon focuses on the "where" rather than the "what." It assumes that the data have been measured and refrains from discussing the techniques of measurement science, other than to recognize that measurement is prone to error. Pure mathematics, even when dealing with vague concepts, provides precise answers that can be verified by anyone. Even statistical analysis uses processes that can be programmed into a computer to give a consistent answer, even when the underlying assumptions are not met or the hypothesis has been incorrectly formulated. The apparent exactness of an answer does not mean that it is correct. To understand the output from, for example, a geographic information system one needs to understand the quality of the data that are entered into the system, the algorithms behind the data processing, and the limitations of the graphic displays.

This book deals with only part of the bigger picture. It focuses on the basic mathematical techniques, building like the whole of mathematics in a series of steps that are the foundations for a deeper understanding. It seeks to lay the foundations for the more complex forms of manipulation that arise in the handling of spatially related data.

The technology behind Geographic Information Systems (GIS) allows such data to be gathered, processed, and displayed. The power and appeal of such systems often lie in their graphical output, the maps that they create. Users of GIS need to understand the quality of that output so that they can advise others on the integrity of their results. The issue is not a matter of which buttons to push but rather of the quality of the information that has been produced. Quality means "fitness for purpose" and "safety in use."

This book therefore looks at some of the fundamentals and provides an introduction to spatial data manipulation through which users of GIS may come to understand whether what they do results in what can genuinely be described as a "quality product." It has been copyedited for an American market, hence the spelling of words such as "meter" for the English "metre" and "center" for "centre."

Contents

List of Tables and Figures xi

Biography xv

Chapter 1.	**Characteristics of Geographic Information** 1
	1.1 Geographic Information and Data 1
	1.2 Categories of Data 2
	1.3 Spatial Referencing 4
	1.4 Lines and Shapes 7

Chapter 2.	**Numbers and Numerical Analysis** 9
	2.1 The Rules of Arithmetic 9
	2.2 The Binary System 13
	2.3 Square Roots 15
	2.4 Indices and Logarithms 17

Chapter 3.	**Algebra — Treating Numbers as Symbols** 23
	3.1 The Theorem of Pythagoras 23
	3.2 The Equations for Intersecting Lines 25
	3.3 Points in Polygons 30
	3.4 The Equation for a Plane 31
	3.5 Further Algebraic Equations 31
	3.6 Functions and Graphs 36
	3.7 Interpolating Intermediate Values 38

Chapter 4.	**The Geometry of Common Shapes** 43
	4.1 Triangles and Circles 43
	4.2 Areas of Triangles 46
	4.3 Centers of a Triangle 49
	4.4 Polygons 51
	4.5 The Sphere and the Ellipse 53
	4.6 Sections of a Cone 55

Chapter 5.	**Plane and Spherical Trigonometry** 59
	5.1 Basic Trigonometric Functions 59
	5.2 Obtuse Angles 63
	5.3 Combined Angles 65
	5.4 Bearings and Distances 67
	5.5 Angles on a Sphere 72

Chapter 6.	**Differential and Integral Calculus**		77
	6.1	The Basis of Differentiation	77
	6.2	Differentiating Trigonometric Functions	81
	6.3	Polynomial Functions	83
	6.4	Basic Integration	85
	6.5	Areas and Volumes	87
Chapter 7.	**Matrices, Determinants, and Vectors**		91
	7.1	Basic Matrix Operations	91
	7.2	Determinants	94
	7.3	Related Matrices	95
	7.4	Applying Matrices	98
	7.5	Rotations and Translations	99
	7.6	Simplifying Matrices	105
	7.7	Vectors	109
Chapter 8.	**Curves and Surfaces**		115
	8.1	Parametric Forms	115
	8.2	The Ellipse	119
	8.3	The Radius of Curvature	121
	8.4	Fitting Curves to Points	122
	8.5	The Bezier Curve	128
Chapter 9.	**Transformations**		131
	9.1	Homogeneous Coordinates	131
	9.2	Rotating an Object	132
	9.3	Hidden Lines and Surfaces	139
	9.4	Map Projections	140
	9.5	Cylindrical Projections	142
	9.6	Azimuthal Projections	145
	9.7	Conical Projections	147
Chapter 10.	**Basic Statistics**		153
	10.1	Probabilities	153
	10.2	Measures of Central Tendency	156
	10.3	The Normal Distribution	160
	10.4	Levels of Significance	163
	10.5	The t-Test	165
	10.6	Analysis of Variance	166
	10.7	The Chi-Squared Test	169
	10.8	The Poisson Distribution	170

Chapter 11.	**Best-Fit Solutions**		173
	11.1	Correlation	173
	11.2	Regression	176
	11.3	Weights	180
	11.4	Linearization	182
	11.5	Least-Square Solutions	184

Further Reading 197

Index 199

List of Tables and Figures

Tables

1.1	Points, Lines, and Areas on Maps	4
3.1	Values of y and x for $y = 0.5x^2$	37
5.1	Signs of Sines and Cosines	65
5.2	Traverse Calculation	71
6.1	Data for $y = 1 + 9x - 6x^2 + x^3$	80
8.1	Data for Points on a Piece-wise Quadratic	125
8.2	Data for a Bezier Curve	129
10.1	Partial Areas Under the Normal Curve	163
10.2	Levels of Significance (P) for Values of t (Given 9 Degrees of Freedom)	166
10.3	Data Classified into Rows and Columns	168
11.1	Framework for Calculating 'r'	174
11.2	Conditions to be Satisfied	184
11.3	The Normal Equations	186
11.4	Weighted Normal Equations	187
11.5	Observations and Conditions	188
11.6	Relationships to be Optimized	188
11.7	The Differentiated Equations	189
11.8	The Relationships between the Correlatives	189
11.9	The Equations for the Correlatives	190
11.10	Solving for the Correlatives	190
11.11	The Correlative Matrix	193

Figures

1.1	A scale bar	3
1.2	Rectangular Cartesian co-ordinates	4
1.3	Nonrectangular or skewed grid	5
1.4	Polar coordinates	5
1.5	Latitude and longitude	6
1.6	Lines as vectors	7
1.7	A line as a raster image	7
1.8	Topology	8
2.1	The distance from A to B	15
3.1	Rotating a triangle	24
3.2	Intersecting lines	25
3.3	The slope of a line	25
3.4	Parallel lines	28
3.5	Clipping to a window	28

3.6	Similar triangles	29
3.7	Point-in-polygon	30
3.8	A plane surface	31
3.9	The equation for a circle	33
3.10	Graph of the function $y = 0.5x^2$	37
3.11	Graph of $y = 1/x$	38
3.12	Linear interpolation and extrapolation	39
3.13	The midpoints of the sides of a triangle	40
3.14	Interpolation of heights down a slope	40
3.15	Interpolating contours between spot heights	41
3.16	Alternative triangulation networks	41
4.1	The angles of a triangle	44
4.2	Angles subtended by arcs	45
4.3	The area of a triangle	46
4.4	The area of a trapezium	46
4.5	The area of a triangle by co-ordinates	48
4.6	The centroid	49
4.7	The orthocenter	49
4.8	The incenter	50
4.9	The circumcenter	50
4.10	Inscribed and circumscribed circles	50
4.11	Straight-line figures	51
4.12	Two ways to divide a quadrilateral	51
4.13	Triangulation networks	52
4.14	Theissen polygons and Delaunay triangles	53
4.15	Great and small circles	54
4.16	Ellipse and auxiliary circle	54
4.17	An ellipse and its directrix	55
4.18	A parabola	56
4.19	A hyperbola with its asymptotes	56
4.20	Sections of a cone-circle and ellipse	57
4.21	Sections of a cone-parabola and hyperbola	57
5.1	Similar right-angled triangles	59
5.2	An altitude of a triangle	61
5.3	Towards a right-angle	63
5.4	A circle with unit radius	63
5.5	Angles in the second quadrant	64
5.6	Angles in (a) third and (b) fourth quadrants	64
5.7	The cycle of values of sin A	65
5.8	Combining adjacent angles	66
5.9	Angle and bearing measurements	67
5.10	Fixing points from observed angles	69
5.11	A traverse	69
5.12	The spherical triangle	72
5.13	Spherical angles	73
5.14	The sine formula for spherical triangles	73

LIST OF TABLES AND FIGURES

5.15	The cosine formula for spherical triangles	74
5.16	Colatitudes	76
6.1	Tangents to a curve	77
6.2	The slope and the normal	78
6.3	A cubic curve	80
6.4	Small angles	81
6.5	Area beneath a curve	87
6.6	Area of an irregular shape	89
6.7	Volumes of a cylinder and cone	90
7.1	Submatrices and minors	97
7.2	Shift or translation of origin	99
7.3	Rotation of axes	100
7.4	A skewed grid	102
7.5	Positive rotations — left hand rule	103
7.6	Photogrammetric rotations	104
7.7	The axes **i**, **j**, and **k** for vector **P**	110
7.8	Vector addition	110
7.9	Direction cosines	111
7.10	Dot and cross products	111
7.11	A parallelepiped	112
7.12	Vectors and a plane	113
8.1	Orthogonal lines	116
8.2	Tangents to a circle and an ellipse	116
8.3	The ellipse	117
8.4	Normals to an ellipse	119
8.5	θ and ø	121
8.6	Radius of curvature	121
8.7	Fitting a second-degree curve	124
8.8	Two quadratics fitted to three points	125
8.9	A piecewise cubic	126
8.10	Looping curve	127
8.11	A Bezier curve with two control points	128
8.12	Two versions of a Bezier curve	129
9.1	Points at infinity	132
9.2	Vanishing points for a rectangular block	132
9.3	A barn	134
9.4	The barn after two rotations	137
9.5	The affine and perspective projections of the barn	138
9.6	Transformation into a perspective view	140
9.7	The elemental triangle	141
9.8	The simple cylindrical projection	141
9.9	Cylinder, cone, and plane	142
9.10	Cylindrical equidistant, equal area and conformal	143
9.11	Elements on a sphere and plane	143
9.12	The Transverse Mercator	144
9.13	Zenithal projections	146

9.14	Oblique azimuthal	147
9.15	Conical projection with 1 or 2 standard parallel	148
9.16	Elemental triangles for conical projections	148
9.17	Conical equidistant with one standard parallel	149
10.1	A plot of equal probability after 15 events	156
10.2	A plot of probability of 'n' events	162
10.3	One- and two-tailed tests	167
11.1	Regression line	176
11.2	Residuals from the regression line	176
11.3	Example of regression line	178
11.4	The braced quadrilateral	191

Biography

Peter Dale trained as a land surveyor and worked for seven years in Uganda before entering the academic world, where he ultimately became Professor in Land Information Management at University College London. He is an Honorary President of the International Federation of Surveyors and was awarded an OBE in recognition of his services to surveying.

CHAPTER **1**

Characteristics of Geographic Information

CONTENTS

1.1 Geographic Information and Data 1
1.2 Categories of Data 2
1.3 Spatial Referencing 4
1.4 Lines and Shapes 7

1.1 GEOGRAPHIC INFORMATION AND DATA

It used to be said that geography was about "maps," as distinct from "chaps." Without doubt today it is about both, and a lot more besides. Ultimately, geography is about making sense of the world around us and this is done by observing, measuring, and processing data about the environment and then presenting the information either as text or pictorially. In particular, it is concerned with why things are where they are.

In recent years, much has been said and written about *geographic information systems* or *GIS*, which are tools that can help the process of understanding. Although there are various interpretations of what is meant by the acronym "GIS," the majority of people would accept that it includes a computer system of hardware and software that can be used to record, manage, integrate, manipulate, analyze, and display data that are spatially referenced to the Earth. The term "spatially referenced" means that their location can be described by measured quantities. *Data* are basic facts that can somehow or other be measured and turned into information.

Information is the commodity that is used by people when they make decisions. Too many facts can be confusing — there may be many different possible routes from one's home to the nearest shopping mall, each of which has its own quality of road surface, slopes, twists, turns and intersections, street lamps, drain covers, etc. All these facts about each route can be measured and recorded, but all the average user really wants to know is the shortest route. This is a piece of information that can be extracted from the basic data.

The term "shortest" is ambiguous since it could mean shortest in terms of time or shortest distance; these are not necessarily the same. The types of data that need to be collected depend on the use to which the information is to be put. The required output determines the required input and the manner in which the data may need to be processed. One can, of course, start with a set of data and see what sense can be made of all the facts and figures. Frequently, the most effective way to do this is through pictures, especially maps and graphs. Advocates of the use of GIS often quote the 19th-century case in London where the locations of cases of cholera were plotted on a map, which then showed clearly that they formed a cluster around an infected well whose water had become contaminated.

When processing data, two golden rules always apply:

1. Bad data plus good processing give rise to unreliable information.
2. Good data plus bad processing also give rise to unreliable information.

If data are to be converted into good-quality information, then both the data and the method by which they are processed must be of good quality, that is, they must be "fit for purpose" or "safe in use." In the discussions that follow, we will focus on the basic principles underlying how data are processed and not on the technical aspects of measurement or how the data are acquired.

1.2 CATEGORIES OF DATA

Data come essentially in two forms — *categorical* and *numerical*. As their name suggests, categorical data are those that are placed in a category or classification system. Such data are sometimes referred to as *nominal data* and have no numerical value. Whether a piece of fruit is an apple or a pear or something else depends on the object itself and the way in which fruits have been classified. For many objects, there are internationally and scientifically recognized standards for classification, although even then there is the occasional dispute over whether some new discovery belongs as a sub-set of an existing class or whether it represents a totally new species.

With some data, categorization is less scientific, for instance when designating the type of land use at a particular location. Although within each country there may be a national land-use classification system, it does not mean that all those who record land use abide by it and it certainly does not follow that every country uses the same system. A building may be used in several different ways with, for example, the basement as a gymnasium, the ground floor as a shop, the next floor as commercial offices, and the top floor as residential accommodation. In spite of national guidelines, investigators may still disagree as to how the use of the building should be categorized. It is, however, not the aim of this book to analyze the problems of data classification but rather to note that it is an issue that intimately affects the quality of data.

Once data have been categorized, they can be subjected to comparison without being quantified. Thus, the data can be placed in a rank so that '*a*' is said to be more than '*b*,' which is more than '*c*,' etc. Such data are described as *ordinal*, an example of which is a list of preferences (area '*a*' is a nicer place to live than area '*b*,' etc.). Various statistical tests exist to process and analyze the differences between ranks or sequences of ordinal data but these also will not be discussed here.

CHARACTERISTICS OF GEOGRAPHIC INFORMATION

Once data have been categorized, it is often necessary to indicate their magnitude. This may be done through the use of *discrete* or *continuous* variables. A discrete variable is one that can only take distinct values, while a continuous variable is one that changes only gradually, allowing any intermediate values.

Some data can only be measured in terms of whole numbers (called *integers* — such as the number of children in a family), while other items can be measured on a continuous scale (such as the height of each child). One can, of course, talk about the average family size being 2.54 children, even though it is impossible to have 0.54 of a child. Such a figure is useful for some practical purposes, especially when associated with an estimate of its reliability, as discussed in Chapter 10.

Discrete variables are precise and are often expressed as whole numbers or integers (0, 1, 2, 3, etc.). More particularly, they can take a succession of distinct values at set intervals along a scale for which there are no intermediate values. Such data are often referred to as *interval data* (Figure 1.1).

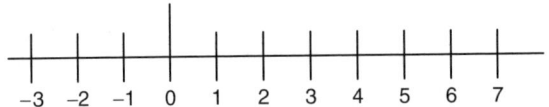

Figure 1.1 A scale bar.

The data may be positive or negative, but the items can only be compared quantitatively on the basis of the differences between them. Only when the values are *absolute* can valid conclusions be drawn about their relative sizes. One can say that a family with four children has twice as many youngsters as a family with two children because "zero children" is an absolute point of reference. One should not, however, say that a temperature of 16°C is twice as hot as a temperature of 8°C as zero on the centigrade scale is an arbitrarily chosen point.

The highest level of measurement is the *ratio scale*, which differs from the interval scale in that it relates to absolute zero (in the case of temperature, this is approximately −273°C). Absolute temperature, length, and breadth are examples of measures on a ratio scale. They are *continuous variables* in that they are not restricted to integer forms but can take any value whatsoever from zero upward. The numerical quantity used to express the measurement of a continuous variable, such as the length of a line or the area of a field, presupposes a standard unit of measure. The numerical value represents the ratio between the quantity measured and the unit of measurement (e.g., the meter or "metre").

Geographical data have one particular characteristic that distinguishes them from all other forms of data, namely location. Graphical data can be plotted on a map and can be represented by points, lines, and areas. From a theoretical perspective, a point on its own has no dimension, a line has one dimension (length), an area has two, and a volume three. In practice, a point on a map is a blob or a very small area, while a line has thickness as well as direction. Each has a category (the "what") representing some attribute or attributes associated with it, and each has a location ("where"). Examples of how points, lines, and areas may be used by cartographers are shown in Table 1.1.

4 INTRODUCTION TO MATHEMATICAL TECHNIQUES USED IN GIS

Table 1.1 Points, Lines, and Areas on Maps

Feature	Points	Lines	Areas
Physical objects	Corner of building	Road network	Planning zone
Statistical values	Sampling point	Isoline	Layer tints
Areas	Central point	Boundary line	Polygon
Surfaces	Heights point	Contour	Hill shading
Text	House numbers	Street names	District names

1.3 SPATIAL REFERENCING

To define the location of any point, there must be some reference to which the point can be related. The most common reference system uses a rectangular grid composed of squares of a standard size. For absolute position (as distinct from the relative position), the grid must have a point of *origin* from which measurements are taken. Points may then be located so far east (or to the right) of the origin and so far north (or up the page), using a standard unit of measure. The two distances are called the co-ordinates or more particularly the *Rectangular Cartesian coordinates* (named after the French mathematician Descartes).

In Figure 1.2, the rectangular grid coordinates of P are (x, y) relative to the origin and are shown in the figure as $(6, 5)$. The idea can be extended to three dimensions by adding the height above the origin. Although in some countries the x-direction is taken as being to the north or upward, here we shall follow the convention that the direction across the page is the x-direction while the direction up the page is described as the y-direction ('x, y', or "in the door and up the stairs"). Height is then in the z-direction. For any point on a three-dimensional object, the coordinates would be (x, y, z). For a normal two-dimensional display, $z=0$.

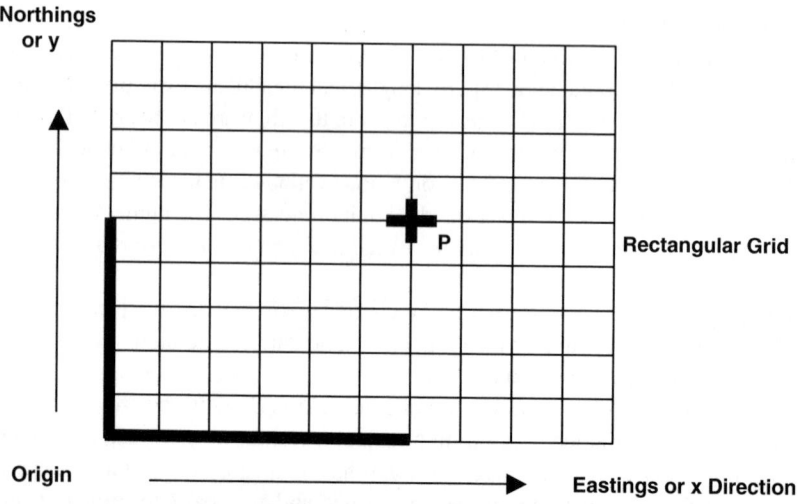

Figure 1.2 Rectangular Cartesian co-ordinates.

CHARACTERISTICS OF GEOGRAPHIC INFORMATION

Simple mathematical techniques can be used to analyze the locations of points that have been referenced to a rectangular grid. Sometimes, it is useful to use a nonrectangular or skewed grid, for example, when trying to show three dimensions on a flat piece of paper (Figure 1.3). Data manipulation is slightly more complicated in these circumstances, although the underlying principles are the same.

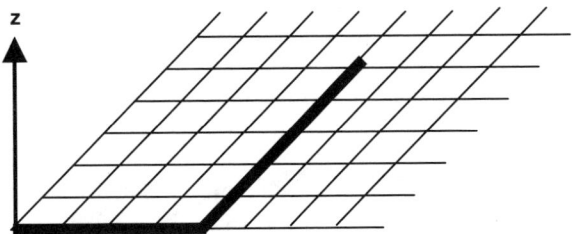

Figure 1.3 Nonrectangular or skewed grid.

An alternative way of measuring the location of a point is through the use of *polar coordinates* (Figure 1.4). These describe points by their distance from an origin and their direction relative to some reference line. The direction is known as the *bearing* and is normally measured clockwise from the north (or up the page). In Figure 1.4, P has polar coordinates (θ, d) relative to the origin and the direction of north, θ being the Greek letter theta.

Angles and distances are examples of measures on the ratio scale. Distances are normally expressed as a ratio, for instance, between the amount of space between two marks that originally designated the length of the international standard metre (now defined in terms of the wavelength of certain light) and the amount of space between two points. As we will show later, trigonometrical formulae allow polar coordinates (bearings and distances) to be converted into Cartesians (eastings and northings or x and y) and *vice versa*.

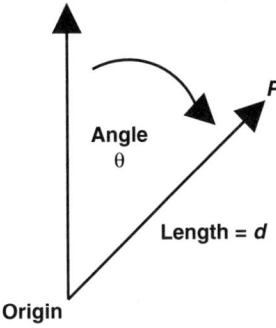

Figure 1.4 Polar coordinates.

Angles are ratios between the amount of turning and a complete turn. They may be expressed either 360 degrees–written as 360° (each degree being subdivided into

60 minutes each with 60 seconds) or 400 grads (where 100 grads equates with a quarter turn, with submeasurements being expressed as decimals) or 2π (two pi) radians, where pi is the ratio between the diameter of a circle and its circumference.

Angular measures are important in surveying where positions may be expressed as if the Earth were a sphere using spherical coordinates. The *latitude* of a point is its angular distance north or south of the equator and is often represented by the Greek letter phi or ø. The *longitude* of a point is an angular measure east or west of the Greenwich Standard Meridian; it is normally represented by the Greek letter "lambda" or λ (see Figure 1.5).

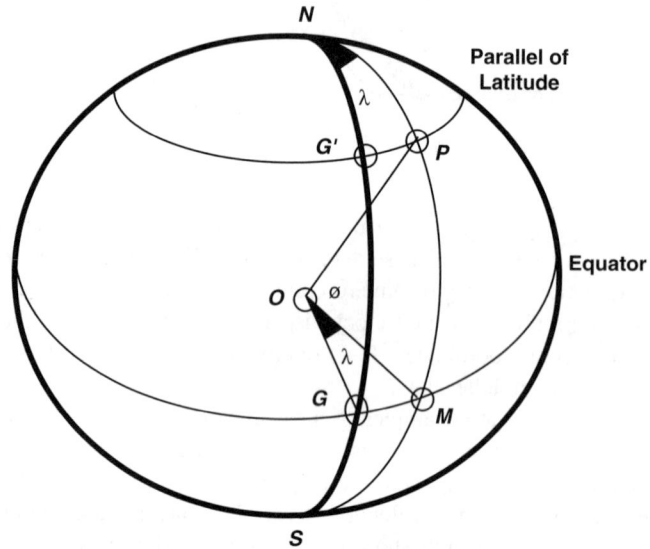

GG' is the Greenwich Meridian of Origin (Zero Longitude)
GM is the Equator (Zero Latitude)
O is the center of the spherical Earth
The angle GOM is the **longitude** of $P = \lambda$
The angle MOP is the **latitude** of $P = ø$

Figure 1.5 Latitude and longitude.

The altitude or height of any point is measured as a distance above a reference surface, which is a mathematical shape that best approximates to the size and shape of the Earth. It is not normally assigned a Greek letter and hence the coordinates of points are expressed as (ø, λ) or as (ø, λ, H).

For more accurate work, the shape of the Earth is assumed to be an *ellipsoid* (an ellipse rotated on its shorter axis) as discussed in Chapter 4; but for many practical purposes, the Earth can be regarded as a sphere. The word *accurate* as used here relates to proximity to the truth. The word *precision* will be used to refer to the exactness with which a value is expressed, whether the value is right or wrong. The word "precision" is often used to describe the number of decimal places that represent the quantity — for example, a distance expressed as 2.105 m (where 'm' means meters) is more precise than 2 m, although the latter may be closer to the truth and hence more accurate.

1.4 LINES AND SHAPES

On a flat surface, a straight line represents the shortest distance between two points. On a curved surface such as a sphere, a so-called straight line bends with the surface, while a line of sight is even more bent because the light is refracted in the atmosphere. We shall not deal with the consequences of the latter effect. In Chapter 9, we will consider how to transform measurements on curved surfaces onto a plane through map projections; for now, we shall focus on a flat Earth and two-dimensional representations.

Geographic information systems handle lines either as vectors or as rasters. A *vector* is a quantity that represents both direction and distance. A polar coordinate as described above is an example of a vector quantity relative to the origin of the coordinate system. Every straight line has a direction and length (such as A to B in Figure 1.6), while a curved line may be considered as being composed of a series of short lines or vectors. In Chapter 7, there is further discussion on vectors.

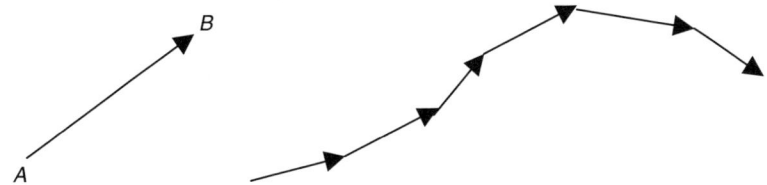

Figure 1.6 Lines as vectors.

On the other hand, a line can, be considered as a series of points adjacent to each other, each point being of small but finite size (Figure 1.7). A television screen or a dot matrix printer produce what may appear like smooth curved lines to the eye but,

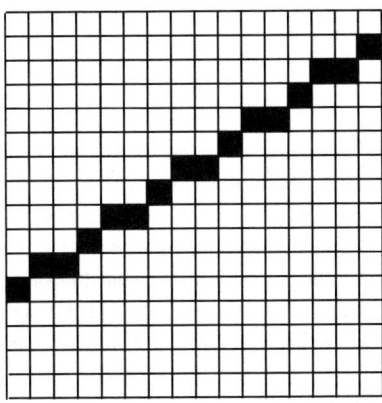

Figure 1.7 A line as a raster image.

which, in practice are a series of points on a grid. Such a representation is called a *raster image* and each small square is known as a *pixel*.

Raster imaging can be used to show lines or areas. Raster data are simple for a computer to handle, but require relatively large amounts of computer storage. Given an area 20 cm by 20 cm and a grid cell size or resolution of 100 dots to the centimetre (i.e., one-tenth of a millimetre), a storage capacity for four million bits of information will be required. For each vector, it is necessary to record only the coordinates of the start and end points of straight line sections.

A line on a flat surface may be regarded as an item in its own right; alternatively, it may be regarded as the division between two areas: one on the left and the other to the right of the line. *Topology* involves the study of *adjacency*, that is, what lies beside a given area, *containment* (what is contained within it) and *connectivity* (how lines or areas are connected to other lines or areas).

Thus, in Figure 1.8, the area B is adjacent to the area A, while the area C is contained within the area A. The point P is connected to the point Q. The fact that the line PQ may be straight or curved is of no significance; what is important is that there is a division and areas A and B lie on opposite sides of it. Whether something belongs to one group or another is often an important consideration, but the mathematics of this will not be discussed here.

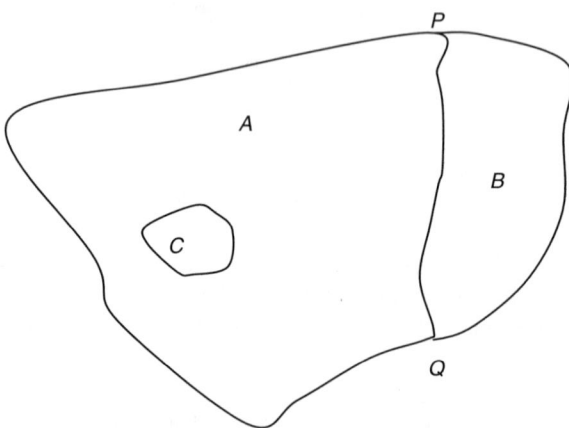

Figure 1.8 Topology.

Throughout this book, the focus will be on two and three dimensions (length, breadth and height or latitude, longitude, and altitude). Time is, of course, a further dimension and strictly speaking we should consider not only the (x, y) or (x, y, z) coordinates of a point but also the (x, y, z, t), where t is time. From a mathematical perspective, there is no inherent reason why we should not consider a world in which there are many more dimensions but this is beyond the scope of the present book. Likewise, we shall neither develop the ideas of topology nor the study of set theory, which is concerned with the relationship between the properties of data sets, although we will review elementary statistical processes in Chapters 10 and 11.

CHAPTER 2

Numbers and Numerical Analysis

CONTENTS

2.1 The Rules of Arithmetic 9
2.2 The Binary System 13
2.3 Square Roots 15
2.4 Indices and Logarithms 17

2.1 THE RULES OF ARITHMETIC

Mathematics is about applying rules, most of which are straightforward. The whole of arithmetic is based essentially on seven axioms, as shown in Box 2.1.

Beyond arithmetic, these axioms may not apply, for instance, when two raindrops running down a window pane come together to make one raindrop so $1 + 1 = 1$. Furthermore, computer programmers often write "$N = N + 1$," meaning "Take the number in the box labeled N, add one to that number and put it back in the box labeled N"; although partially an arithmetic operation, the use of the '=' sign has a different meaning from what we are considering here. With respect to the sequence of operations, if you rotate a dice forward and then sideways, it will land in a different position than if you rotate it sideways and then forward. This illustrates how in some circumstances the sequence of operations can be important, and we shall discuss this further in Chapter 7 within the context of matrices.

In this chapter, we shall deal with simple arithmetic, for which the axioms in Box 2.1 are fundamental. They are all necessary but not quite sufficient. Consider the calculation $2 + 3*4$. A pocket calculator will show this as $2 + 3 = 5$; then enter 4 and multiply to give the answer 20. On the other hand, $3*4 = 12$. Add $2 = 14$. The same sum carried out in a different order gives a different answer.

Hence, we must have rules of priority. The simplest way to handle this is to place brackets round the groups that are together. Thus, in the first case, we have $(2 + 3)*4$, while in the second case we have calculated $2 + (3*4)$. We must distinguish between these two cases. This leads to Rule 1 in Box 2.2.

9

Box 2.1 — Axioms of Arithmetic

Axiom 1.

For any numbers 'a' and 'b',
  a plus b has the same value as b plus a.
Also, a times b has the same value as b times a.

In symbols, $a + b = b + a$

  where '+' means "add"

  $a*b = b*a$

  where '*' means "multiply"

The latter may also be written as $ab = ba$

Axiom 2.

For any three numbers a, b, c

$$(a + b) + c = a + (b + c)$$

Also, $(ab)c = a(bc)$

Axiom 3.

For any numbers a, b, c

$$a*(b + c) = a*b + a*c$$

or $a(b + c) = ab + ac$

Axiom 4.

There is a number called **zero** (0) such that for any 'a'

$$a + 0 = a$$

(Historically, there was a debate as to whether 0 was actually a number).

Axiom 5.

There is also a number 1 such that $a*1 = a$

Axiom 6.

For every value of 'a', there is a number 'd' such that

$$a + d = 0$$

(This then introduces the whole range of negative numbers)

Axiom 7.

Provided 'c' does NOT equal zero ($c \neq 0$) then

if $c*a = c*b$ then $a = b$

(Similarly, if $a + c = b + c$, then $a = b$, although this also applies if $c = 0$ but not when c is infinitely large)

NUMBERS AND NUMERICAL ANALYSIS

> **Box 2.2 — Rule 1**
>
> Place items that are together in brackets and deal with what is inside the brackets first.

Where there are no brackets, we have to decide which comes first — addition (+), subtraction (−), multiplication (∗), or division (/). In fact, it does not normally matter whether we multiply and then divide in that order or divide first and then multiply. Thus,

$$3*4/2 = (3*4)/2 = 12/2 = 6,$$

while $3*(4/2) = 3*2$, which also equals 6

The same occurs with addition and subtraction. Thus,

$$2 + 3 - 4 = (2 + 3) - 4 = 5 - 4 = 1,$$

while $\qquad 2 + (3 - 4) = 2 - 1$, which also equals 1

The key point is that we should do the multiplication or division before the addition or subtraction. The second rule is therefore as shown in Box 2.3.

> **Box 2.3 — Rule 2**
>
> If there are no brackets or when what is inside the brackets has been evaluated, then deal with multiplication or division before addition or subtraction.

Numbers may be positive or negative and when handling these, simple rules also apply. Thus, adding a negative number is the same as subtracting a positive while subtracting a negative number is the same as adding a positive.

Put in another way,

$$4 + (-3) = 4 - 3 = 1,$$
while $\qquad 4 - (-3) = 4 + 3 = 7$

Multiplication and division follow Rule 3 in Box 2.4.

> **Box 2.4 — Rule 3**
>
> (positive)∗(positive) = + = (positive)/(positive)
> (positive)∗(negative) = − = (positive)/(negative)
> (negative)∗(positive) = − = (negative)/(positive)
> (negative)∗(negative) = + = (negative)/(negative)

Ordinary numbers not only come as positive or negative but also as integers or real numbers. An *integer* is a whole number such as 1, 2, 3, or 4. *Real* numbers basically occur in between and are often expressed in *decimal* parts (i.e., in submultiples of 10). Real numbers are either rational or irrational. A *rational* number is one that can be expressed as the ratio of two integers, for example, 1.125 = 1 + (125/1000). Irrational numbers are ones that cannot be expressed in this way such as π, which is the ratio between the circumference of a circle and its diameter. π has been calculated to millions of decimal places, with the sequence of numbers never being repeated.

A *fraction* is a number that can be expressed in the form of two integers such as (*I/J*), where *I* and *J* are whole numbers; for example, (1/3), (5/8), and (13/25) are all fractions. *I* is known as the *numerator* (the number that indicates how many there are) and *J* is the *denominator* (giving the denomination or category of the fraction). Fractions can be converted into decimal parts simply by dividing the denominator into the numerator; thus, 5 and 8 are integers, while 5/8 is a fraction that can be expressed as the decimal number 0.625.

Numbers can also be *imaginary* in that they obey special rules that contradict the above, notably that when multiplied by themselves they can produce a negative answer. They are particularly useful when handling vectors (which will be discussed in Chapter 7), but will not be dealt with here. We shall treat all numbers as integers or as real numbers.

Many integers are the product of smaller integers, each of which is a *factor* of the bigger number. Thus, the factors of 14 are two and seven since (2∗7) = 14. Positive numbers that cannot be expressed as the product of at least two smaller integers are said to be *prime* numbers. Thus, 1, 2, 3, 5, 7, 11, 13, 17, 19, 23, 29, 31, 37, 41, 43, and 47 are all the prime numbers below 50.

Factorizing is important when dealing with fractions as it permits simplification. By removing all the common factors from the numerator and denominator, we can, for example, simplify $\frac{561}{2431}$ to $\frac{187*3}{187*13}$, which reduces to $\frac{3}{13}$. Finding the *highest common factor* (here 187) can greatly simplify subsequent data processing.

Conversely, we may wish to combine two fractions such as adding $\frac{7}{16}$ to $\frac{13}{40}$. Here, we need to find the *lowest common denominator*, that is, the smallest number into which the two denominators can divide. In this case, it would be 80 since it is the smallest integer number that can be divided by both 16 and 40. There are, of course, higher numbers such as (16∗40 or 640), but 80 is the smallest number that is divisible by both. Hence, by expressing the first fraction $\frac{7}{16}$ as $\frac{5*7}{5*16}$ or $\frac{35}{80}$ and the second as $\frac{2*13}{2*40}$ or $\frac{26}{80}$, the two fractions can be added (Box 2.5) $\frac{7}{16} + \frac{13}{40} = \frac{35}{80} + \frac{26}{80} = \frac{61}{80}$.

Box 2.5 — Rule 4

To add or subtract fractions, they must share a common denominator.

Alternatively, fractions can be expressed as decimal numbers and then processed accordingly. A decimal number consists of an integer, a decimal point, and a series of fractions of the number 10. In some countries, notably in parts of mainland Europe, a comma is used instead of a decimal point, while a full stop is used to separate out thousands. Thus, one million plus twenty-three hundredths would be written as 1.000.000,23 but in this text we shall use the comma to separate the thousands and a full stop as the decimal separator so that our number will be shown as 1,000,000.23.

Decimals are easier to add or subtract than fractions. Thus, in the above example, $\frac{7}{16} = 0.4375$ while $\frac{13}{40} = 0.3250$, yielding a sum of 0.7625, which is the same as $\frac{61}{80}$ expressed as a decimal.

Decimal numbers use a sequence of numerals that were derived from the Arabic system, representing 0 through 9. Thus, the number 8642 represents

$$8*(10*10*10) + 6*(10*10) + 4*(10) + 2*1$$

By using indices as discussed later, we can write $10*10*10$ as 10^3, $10*10$ as 10^2, 10 as 10^1, and even 1 as 10^0.

This means that $8642 = 8*10^3 + 6*10^2 + 4*10^1 + 2*10^0$.

Continuing the sequence, $(1/10) = 10^{-1}$, $(1/100) = 10^{-2}$, etc., so that 0.325, for example, represents $3*(1/10) + 2*(1/100) + 5*(1/1000)$, which may be written as

$$0.325 = 3*10^{-1} + 2*10^{-2} + 5*10^{-3}$$

2.2 THE BINARY SYSTEM

Although the decimal numbering system is convenient for human beings, there are useful alternatives. The most common is the binary system, because it is appropriate for computers. The binary system works only with the numbers 0 and 1 (*binary digits* or *bits*) and increases by multiples of 2 rather than 10 (Box 2.6).

Box 2.6 — The Binary System

2 in decimals = 10 = $1*2^1 + 0*2^0$ in binary
3 in decimals = 11 = $1*2^1 + 1*2^0$ in binary
4 in decimals = 100 = $1*2^2 + 0*2^1 + 0*2^0$
8 in decimals = 1000 = $1*2^3 + 0*2^2 + 0*2^1 + 0*2^0$
12 in decimals = 1100 = $1*2^3 + 1*2^2 + 0*2^1 + 0*2^0$
14 in decimals = 1110 = $1*2^3 + 1*2^2 + 1*2^1 + 0*2^0$
16 in decimals = 10000 = $1*2^4$
31 in decimals = 11111 = $2^4 + 2^3 + 2^2 + 2^1 + 2^0$
32 in decimals = 100000 = $1*2^5 + 0$, etc.
255 in decimals = 11111111, which consists of 8 bits.

Early computers used to store numbers in 8 bits, a "bit" being simply "0 or 1," "off or on," or "no charge or with charge." Each storage box within the computer system could hold 8 of these at a time, giving any number between 0 and 255. Now, they may use at least 64 bits. 2^{64} represents a very large number.

In the binary system, the rules for addition, subtraction, multiplication, and division are similar to those for decimal numbers. For example, in decimals, when adding, work from the right to the left and "carry one" if beyond the base number 10 (Box 2.7).

Box 2.7 — Decimal (Base 10) addition

To add 83 + 649, $649 = 6*10^2 + 4*10^1 + 9*1$
Then to add $+83 = 0*10^2 + 8*10^1 + 3*1$
Start on the right (singles) column. Add (=**12**). Retain **2** and carry one over to the tens column. Add the tens (=**13**). Retain the **3** and carry one over to the hundreds column, finishing with
$$\text{Total} = 7*10^2 + 3*10 + 2*1 = 732$$

The same procedure is followed with the binary system where increments go from 1 to 2 (=10) to 4 (=100) to 8 (=1000) to 16 (=10000), etc., doubling each time and adding a zero, as shown in Box 2.8.

Box 2.8 — Binary (Base 2) addition

To add the binary numbers **111 + 101**:
Using **bold** for binary numbers
Given: **111** (in decimals $1*2^2 + 1*2^1 + 1*2^0 = 7$)
Add   **101** (in decimals $\underline{1*2^2 + 0*2^1 + 1*2^0 = 5}$)

Start on the right (singles) column

$$\mathbf{1+1} = \mathbf{10} \text{ (1 lot of } 2^1 \text{ plus } \mathbf{0} \text{ lot of } 2^0\text{)}$$
$$= + \mathbf{0}*2^0 \text{ (carry one lot of } 2^1\text{)}$$

Repeat on the middle (twos) column

$$(\mathbf{1+0}) + (\mathbf{1} \text{ carried over}) = \mathbf{1+1} = \mathbf{10} \text{ (1 lot of } 2^2 + \mathbf{0} \text{ lot of } 2^1\text{)}$$
$$= + \mathbf{0}*2^1 + \mathbf{0}*2^0 \text{ (carry } \mathbf{1}*2^2\text{)}$$

Repeat on the left (fours) column

$$(\mathbf{1+1}) + \mathbf{1} = \mathbf{10} + \mathbf{1} = \mathbf{11} \text{ (1 lot of } 2^3 \text{ plus } \mathbf{1} \text{ lot of } 2^2\text{)}$$
$$= + \mathbf{1}*2^2 + \mathbf{0}*2^1 + \mathbf{0}*2^0 \text{ (carry } \mathbf{1}*2^3\text{)}$$

Add the **1** lot of 2^3 carried over

Total $= \mathbf{1}*2^3 + \mathbf{1}*2^2 + \mathbf{0}*2^1 + \mathbf{0}*2^0$

Hence,   **111 + 101 = 1100** (= 12 in decimals)

Similar procedures are followed for subtraction, multiplication, and division. A pocket calculator, for example, turns the decimal number that is typed on the keypad into a binary number, processes it in that form, and then returns the answer in decimals (which is why sometimes answers such as 79.99999999999999 appear instead of 80). Although numbers such as 11010101001010101 may be ideal for computers, they are tedious for human beings; thus, in what follows, we shall stick to the decimal system.

2.3 SQUARE ROOTS

A number that is multiplied by itself is said to be the *square* of that number. 3*3 is the square of three (since it is the area of a square that is three units by three units) and is usually written as $3^2 = 9$. Thus, 9 is the square of three, while 3 is the square root of 9, which is normally written as $\sqrt{9}$. Similarly, a number may be cubed and have a cube root, for instance, $4^3 = 64$, while $\sqrt[3]{64} = 4$.

Squares and square roots are of particular importance when dealing with coordinate systems. According to the theorem of Pythagoras (a proof of which is given in Chapter 3), in any right-angled triangle, the square of the diagonal length is equal to the sum of the squares on the two shorter sides. Thus, if we have the rectangular Cartesian coordinates for two points A and B and the difference between them in the x-direction or Eastings is 'E' and the difference in the y-direction or Northings is 'N', then the square of the distance from A to $B = E*E + N*N = (E^2 + N^2)$. Put more simply, $AB = \sqrt{(E^2 + N^2)}$ (Figure 2.1).

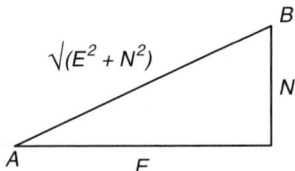

Figure 2.1 The distance from A to B.

In the very special case when $E = 3$ units and $N = 4$ units (or *vice versa*), AB exactly equals 5 since $3^2 + 4^2 = 25 = 5^2$. A similar relationship exists where $5^2 + 12^2 = 169 = 13^2$. Normally, however, the square root will be an irrational number and not an integer. Some numbers are *perfect squares* (1, 4, 9, 16, etc.), in that their square roots are exact numbers. Most numbers do not have an exact square root; hence, any system that generates square roots can only produce an answer that is sufficient for the needs at hand.

The easiest way to find the square root of a number is to use a pocket calculator! If nothing other than a pencil and paper is available, then the value can be obtained by repeated attempts at long division until an answer to the appropriate level of significant numbers is achieved. It is done by *iteration* in which one gets closer and closer to the answer until sufficient significant figures have been reached.

The example given in Box 2.9 shows the calculation of a square root using multiplication and division. It finishes with a number that is an approximation to the true value that, like $\sqrt{2}$, is irrational and the numbers after the decimal point could go on

> **Box 2.9 — Finding a Square Root by Iteration**
>
> 1. The first stage is to mark off the number into pairs.
> Thus, a number such as 27392834 becomes 27, 39, 28, 34
>
> 2. Take the nearest square root of the first pair of significant figures (here 27). Here, the best estimate is 5 (since 5∗5 = 25). Add zeros for the remaining pairs. The first trial number is 5000
>
> 3. Calculate (1/2)∗(Trial + number/trial)
>
> $$= 0.5*(5000 + 27392834/5000)$$
> $$= 0.5*(5000 + 5478.567) = 5239.284$$
>
> 4. Repeat
>
> $$= 0.5*(5239.284 + 27392834/5239.284)$$
> $$= 0.5*(5239.284 + 5228.354) = 5233.819$$
>
> 5. Repeat
>
> $$= 0.5*(5233.819 + 27392834/5233.819)$$
> $$= 0.5*(5233.819 + 5233.814) = 5233.816$$
>
> To three decimal places, 5233.816 is as close to the square root of our original number as we can get. The solution has thus converged quite rapidly from 5000 through 5239 through 5233.82 to 5233.816.

forever. We therefore need to define a suitable level of accuracy and precision beyond which there is no merit in going. A similar problem arises in geomatics where although the calculations may be precise and accurate, even when handling uncertainty, the data that are fed into formulae may be unreliable. The fact that a distance may be calculated in meters and expressed as 123.165284 does not mean that it is accurate to one-thousandth part of a millimeter. Such a number may be calculated to six *decimal places* (six figures after the decimal point) but it does not mean that it is reliable in terms of what is really on the ground to that degree of precision.

The extent to which such a number is reliable can be expressed in terms of its *significant figures*. This is the number of figures after which we can write zeros for all the difference it will make. If the measurement is only reliable to the nearest 10 m, we could say that rather than being 123.165284, the distance is 120 m and is reliable to two significant figures; to the nearest five significant figures, the distance would be expressed as 123.17. In this case, there has been *rounding up* in that the fifth figure (here the number 6 in ".165284") has become 7 and the subsequent numbers have been ignored. The number 123.17 is closer to 123.165284 than is 123.16. If the original figure had been 123.164284, then it would have been *rounded down* and the best estimate of the number to five significant figures would have been 123.16.

NUMBERS AND NUMERICAL ANALYSIS

There is, of course, a potential ambiguity when, for example, rounding 123.5 to three significant figures. Should the answer be 123 or 124? Some people would say "always round down," some "always round up," while others would opt for the answer that is an even number because it can be halved. The convention used here will be that (0, 1, 2, 3, 4) round down to 0, while (5, 6, 7, 8, 9) round up to 10. Thus, the number 123.5 becomes 124.

2.4 INDICES AND LOGARITHMS

Long-hand calculations are tedious but there is a method other than that shown in Box 2.9, whereby we can approach the problem of calculating square roots. It is based on indices. When we write the square of the number 'a' as a^2, the number '2' is said to be the *index* of 'a' or that 'a' has been raised to the *power* of 2. The cube of 'a' is a^3, with 'a' being raised to the power of 3. If we multiply 'a' by itself five times, then we obtain a^5 — for example, 2^5 or "two to the power of 5" equals $2*2*2*2*2$ or 32. For any number 'a' then if its square ($a*a$) is multiplied by its cube ($a*a*a$) we obtain $a*a*a*a*a = a^2*a^3 = a^5$. Note that $5 = 2 + 3$. As a general rule, we add the indices when multiplying the same base number raised to separate powers (Box 2.10).

Box 2.10 — Rule 5

When multiplying any number raised to the power 'm' (a^m) by itself raised to the power 'n' (a^n) then $a^m*a^n = a^{(m+n)}$.

If we divide 'a' to the fifth by 'a' squared, that is, $(a*a*a*a*a)/(a*a)$ or a^5/a^2, then the answer is $(a*a*a)$ or a^3. Here, $5 - 2 = 3$. Rule 5 applies if n is negative as well as if it is positive. Hence, we can write a^{-n} to mean "divide by a^{+n}" or to mean $1/(a^n)$. A particular consequence of this is that if we divided a^n by itself then

$$(a^n/a^n) = a^n*a^{-n} = a^{(n-n)} = a^0 = 1$$

Similarly, if we divide $a^{(n+1)}$ by a^n then (Box 2.11)

$$(a^{(n+1)}/(a^n)) = a^{(n+1)}*a^{-n} = a^{(n+1-n)} = a^1 = a$$

Box 2.11 — Rule 6

Any number raised to the power of zero equals 1 $a^0 = 1$.
Any number raised to the power of 1 is itself $a^1 = a$.

The power of a number does not have to be an integer. Thus, for example,

$$a^{0.5}*a^{0.5} = a^{(0.5 + 0.5)} = a^1 = a$$

or, put another way, the square root of 'a' $= a^{0.5}$.

Thus, the square root of the number 10 can be expressed either as $\sqrt{10}$ or as $10^{0.5}$. Rule 5 applies to all values of '*m*' and '*n*,' not just integers.

This simple rule is the basis of a system of multiplication and division known as *logarithms*. The logarithm of any number '*n*' is the *power* of a fixed number — called the *base* of the system — that gives the same quantitative value as '*n*.' Whereas any base can be used, the most common system of logarithms uses the decimal number 10 as the base. Logarithms to the base 10 are known as common logarithms and written as "log" or $\log_{10}$.

For many scientific purposes, a different base is used, namely the number '*e*' (also called the Euler number after the 18th-century Swiss-born mathematician Leonhard Euler), which has a decimal value that is approximately equal to 2.7182818285.

Logarithms to the base '*e*' are known as *Natural* or *Naperian* logarithms after the 16th-century scientist and mathematician John Napier and written as "ln" or as $\log_e$. '*e*' occurs in a variety of mathematical formulae relating to natural phenomena (hence the name "natural logarithms").

The number '*e*' = 1 + 1/1! + 1/2! + 1/3! + 1/4! + ..., etc., where

2! (spoken as "two *factorial*") = 2∗1 = 2

3! = 3∗2∗1 = 6

4! = 4∗3∗2∗1 = 24, etc.

In general, we have '*n*' factorial (where '*n*' is an integer)

n! = 1∗2∗3∗ ... ∗(*n*-2)∗(*n*-1)∗*n*

For the present, however, we will focus on common logarithms or "logs." These use the base number 10.

Numbers between 1 and 10 have a logarithmic value between 0 and 1; between 10 and 100, the logarithmic value will be between 1 and 2, etc. The value for other numbers can either be calculated, for instance, using a pocket calculator, or obtained from special tables. These will show, for example, that the value for the number 5 is given as 0.6989700. Thus, the logarithm of 5 = $\log_{10}(5)$ = 0.6989700. Put another way, 5 = $10^{0.6989700}$.

Since 50 = 10∗5 = $10^1 \ast 10^{0.6989700}$ then 50 = $10^{1.6989700}$

Thus, the log of 50 = $\log_{10}(50)$ = 1.6989700.

Similarly, the log of 500 = $\log_{10}(500)$ = 2.6989700, etc.

As an example of the use of logarithms, consider 50^2 = 2500

50 = $10^{1.6989700}$ and the logarithm of 50 = 1.6989700

50∗50 = $10^{1.6989700} \ast 10^{1.6989700}$ = $10^{1.6989700+1.6989700}$

= $10^{3.3979400}$ by the law of indices

log 50 + log 50 = 1.6989700 + 1.6989700 = 3.3979400

This is the logarithm of the number 2500. Thus, adding logarithms gives the same answer as multiplying the original numbers.

As another example, the logarithm of the square root of 50 = half of log 50 = 0.849485. This represents the number $10^{0.849485}$, which is the logarithm of the decimal number 7.071068. Thus, the square root of 50 = 7.071.

As a further example, log 100 = 2 since 100 = 10^2. log 50 = 1.6989700

To divide two numbers, we subtract their logarithms so that:

Log (100/50) = Log 100 − log 50 = (2 − 1.698700) = 0.3010300

This represents $10^{0.3010300}$ and is the logarithm of the number '2'.

For numbers less than 1 and greater than 0, the power will be less than zero (see Box 2.12). For instance, 0.5 = $5*10^{-1}$ while 0.05 = $5*10^{-2}$. Hence log (0.5) = −1 + 0.6989700. This can be expressed in one of three ways: either

log (0.5) = −0.3010300

Box 2.12 — Logarithms to the Base 10

$10^0 = 1$ log (1) = 0
$10^1 = 10$ log (10) = 1
$10^2 = 100$ log (100) = 2
$10^3 = 1000$ log (1,000) = 3
$10^4 = 10000$ log (10,000) = 4
$10^5 = 100000$ log (100,000) = 5, etc.

Likewise

$10^{-1} = 0.1$ log (0.1) = −1
$10^{-2} = 0.01$ log (0.01) = −2
$10^{-3} = 0.001$ log (0.001) = −3
$10^{-4} = 0.0001$ log (0.0001) = −4
$10^{-5} = 0.00001$ log (0.00001) = −5, etc.

There is no real value for log(0) — it would be minus infinity

---

or we can separate the substantive number from the decimal point and write

log (0.5) = $\overline{1}$.6989700 (read as "bar one plus 0.6989700")

Alternatively, we can add ten and call it 9.6989700, knowing that for most calculations in geomatics we are not dealing with numbers as large as 10^{10}. So

log (0.5) = 9.6989700

log (0.05) would be either −1.3010300 or else bar 2 plus 0.6989700 (= $\overline{2}$.698970) or 8.698970.

The advantage of the second and third systems is that all logarithms are positive numbers to the right of the decimal point and all multiplications can be done by addition. On the other hand, most pocket calculators use the negative number approach.

Note that a negative logarithm represents a positive number between 0 and 1. There can be no such thing as the logarithm of a negative number since plus and minus relate to multiplication and division and the location of the resulting decimal point.

Box 2.13 — Rule 7

To use logarithms, all numbers must be treated as positive. There is no such thing as the logarithm of a negative number.

Adding two logarithms means that we are multiplying together two numbers; subtracting two logarithms means that we are dividing.

Box 2.14 — Rule 8

When using logarithms (common or natural),

(a) The log of $a*b = \log(a*b) = \log a + \log b$

(b) The log of $a/b = \log(a/b) = \log a - \log b$

(c) The log of $a^n = \log(a^n) = n*\log(a)$

One consequence of Rule 8(c) (see Box 2.14) is that

$$\log_{10}(e^y) = y*\log_{10}e.$$

Also,

$$\log_e(e^y) = y \log_e(e)$$

Now, $\log_e(e) = 1$ just as $\log_{10}(10) = 1$.

Hence,

$$\log_e(e^y) = y \log_e(e) = y$$

As a result,

$$\log_{10}(e^y) = \log_e(e^y)*\log_{10}(e)$$

Replacing e^y by a,

$$\log_{10}(a) = \log_e(a)*\log_{10}(e)$$

'e' has the value 2.718 ... and hence we can calculate the value of $\log_{10}(e)$. This number is known as the *modulus* of common logarithms.

$$\log_{10}(e) = 0.4342944819$$

Thus,

$$\log_{10}(a) = 0.434294482 * \log_e(a)$$

This applies for any positive value of 'a'.

Some calculations become cumbersome when using logarithms. In Box 2.15, we apply logarithms to calculate the distance between two points. The process is tedious, especially when pocket calculators are at hand, because it means constantly switching between logarithms and their inverse (known as *antilogarithms*). Their use is, of course, also dependent on access to tables of logarithms, known as "log tables."

Box 2.15 — Calculating distances using logarithms

Consider $\sqrt{(1243.18^2 + 656.91^2)}$
To square 1243.18, we must multiply it by itself.

Log (1243.18) = 3.0945340 (using 7 figure logarithms)

Hence, the log of $1243.18^2 = 2 * \text{Log}\ (1243.18) = 6.1890680$.
This represents the logarithm of the number 1,545,496

Similarly, Log $(656.91^2) = 2 * 2.8175059 = 5.6350118$.
This represents the logarithm of the number 431530.8.

Using seven-figure logarithms, we have

$(1243.18^2 + 656.91^2) = (1545496 + 431531) = 1,977,027$

Now,   Log (1977027) = 6.2960126

To take the square root, we halve the logarithm

Half Log (1977027) = 3.1480063

To two decimal places, this is the logarithm of 1406.07.

Hence, $\sqrt{(1243.18^2 + 656.91^2)} = 1406.07$

Logarithms are most effective when there are several multiplications or divisions or when dealing with certain statistical distributions, an example of which is given in Chapter 11. They also appear in integral calculus, as we shall see in Chapter 6.

CHAPTER 3

Algebra — Treating Numbers as Symbols

CONTENTS

3.1 The Theorem of Pythagoras 23
3.2 The Equations for Intersecting Lines 25
3.3 Points in Polygons 30
3.4 The Equation for a Plane 31
3.5 Further Algebraic Equations 31
3.6 Functions and Graphs 36
3.7 Interpolating Intermediate Values 38

3.1 THE THEOREM OF PYTHAGORAS

We can apply the ideas of arithmetic to unknown numbers. Algebra is the body of mathematical knowledge that deals with symbols. It deals with constants (such as the number e=2.718..... mentioned in Chapter 2) and variables, such as (x, y), that can be the coordinates of a point, with 'x' and 'y' taking any value. Each value of x and y is a number that can be added, subtracted, multiplied, or divided and hence algebra may be regarded as generalized arithmetic. Algebra is about treating and manipulating numbers as symbols.

The proof of Pythagoras's Theorem given in Box 3.1 applies a number of operations that were discussed in Chapter 2, namely

1. $a^m * a^n = a^{(m+n)}$.
2. $a^m / a^n = a^m * a^{-n} = a^{(m-n)}$.
3. $(a^m)^n = a^{(m*n)}$.

The proof also introduces the use of brackets so that

$$(a + b)*(c + d) = a*(c + d) + b*(c + d)$$

or $(a + b)*(c + d) = (a + b)*c + (a + b)*d$

which in both cases is equal to $a*c + a*d + b*c + b*d$. It also uses indices so that

$$(x + y)^1 * (x + y)^1 = (x+y)^{(1+1)}$$
$$= (x + y)^2 = x^2 + y^2 + 2xy$$

It also makes the point that we can often dispense with the "multiply" and "divide" symbols by writing the above as

$$(a + b)(c + d) = ac + ad + bc + bd$$

Using indices, we can also write $\dfrac{(a + b)}{(c + d)} = (a + b)/(c + d)$ or as $(a + b)*(c + d)^{-1}$, using the -1 index to mean "divide by."

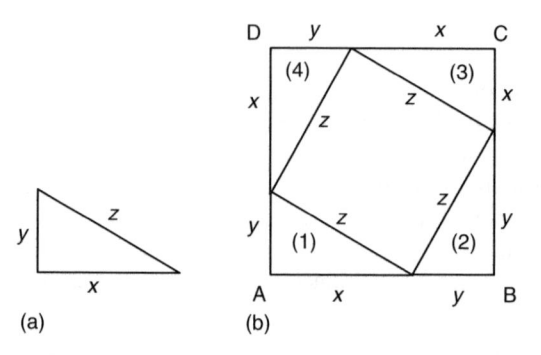

Figure 3.1 Rotating a triangle.

Box 3.1 — One Proof of Pythagoras's Theorem

Consider a simple right-angled triangle with sides of length x, y, and z (the hypotenuse) as in Figure 3.1(a). Copy the triangle three times and rotate it each time to form Figure 3.1b. This builds a square on the hypotenuse or long side.

The triangle in Figure 3.1(a) is half a rectangle with sides x and y. Hence, its area is half of x times y or $1/2\ xy$.

The area of the outer square in Figure 3.1(b) is $(x + y)*(x + y) = (x + y)^2$. We can write this as $x*(x + y) + y*(x + y)$ or as $x*x + x*y + y*x + y*y$ or as $x^2 + y^2 + 2xy$ using the axioms and rules laid down in Chapter 2.

The area of the outer square $ABCD$ is $x^2 + y^2 + 2xy$. It also equals 4 triangles plus one smaller inner square whose sides are of length z. 4 times "$1/2\ xy$" ($= 2xy$) plus the inner square area of $z*z$ gives a total area, $2xy + z^2$. Thus, $2xy + z^2 = x^2 + y^2 + 2xy$, or, removing $2xy$ from both sides of the equation, $z^2 = x^2 + y^2$.

The square on the hypotenuse is equal to the sum of the squares on the other two sides.

Pythagoras deals with the sum of two squares. A particular quadratic form that is often important is known as the "*difference of two squares.*" The expression $(x^2 - y^2)$

ALGEBRA — TREATING NUMBERS AS SYMBOLS

can be *factorized* (broken into factors) into the form $(x - y)(x + y)$. Multiplying these two factors together yields

$$(x - y)(x + y) = x^2 + xy - xy - y^2 = (x^2 - y^2)$$

We shall have cause to use this on a number of occasions.

3.2 THE EQUATIONS FOR INTERSECTING LINES

The rules repeated above provide ample opportunities to solve problems. Consider a rectangular coordinate system with two lines A to B and C to D that intersect at P (as in Figure 3.2).

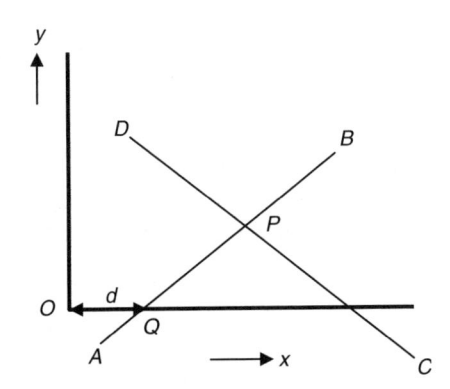

Figure 3.2 Intersecting lines.

The *origin* of the coordinate system is at the point O and the coordinates are

$$A = (x_A, y_A), B = (x_B, y_B), C = (x_C, y_C), \text{ and } D = (x_D, y_D)$$

Let the line AB cut the x-axis at the point Q $(d, 0)$. 'd' is said to be the *intercept* on the x-axis. (Note that superscripts are used as indices, and subscripts such as the '$_A$' in x_A are used as identifiers.)

Consider first a straight line through the origin (Figure 3.3(a)). If the coordinates of any point on this line are (x, y), then the ratio between y and x (or y/x) will always be the same. This is the slope of the line — let us call it 'm.' We then have $y = mx$ for all points on the line, with 'm' a constant number and x and y variables.

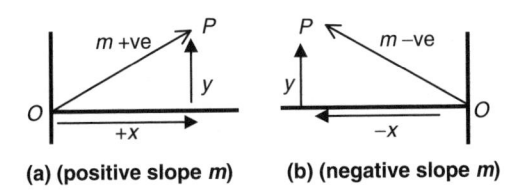

(a) (positive slope m)   (b) (negative slope m)

Figure 3.3 The slope of a line.

If the line slopes backwards (so that as y increases, x decreases and *vice versa*), then '*m*' will be negative (Figure 3.3(b)).

Lines do not necessarily have to pass through the origin but they can always be made to do so by changing the position of the origin. Thus, in Figure 3.2, if the origin were at the point Q where the line AB crosses the x-axis (which is the line where $y = 0$), then AB will pass through this *false origin*. To achieve this, all x values must be reduced by the amount $OQ = $ '*d*.'

Thus, we can express the equation of the line AB using the original coordinates in the form $y = m(x - d)$ or more generally $y = mx + c$ where $c = -m*d$. '*m*' and '*c*' are constants for the line. In fact, we can now describe both lines,

the first (AB) as $\quad y = m_1 x + c_1$

the second (CD) as $\quad y = m_2 x + c_2$

When we know '*m*' and '*c*,' we can calculate the value of '*y*' given '*x*' or *vice versa*. The line crosses the vertical y-axis where

$$x = 0 \text{ and } y = c$$

It crosses the x-axis, where $y = 0$. At this point $x = -c/m$.

We can use the coordinate values to calculate '*m*' and '*c*.' Since the line AB passes through (x_A, y_A) and (x_B, y_B), these values must satisfy the conditions

$$y_A = m_1 x_A + c_1, \quad y_B = m_1 x_B + c_1$$

We can subtract the lower numbers from the upper so that

$$y_A - y_B = (m_1 x_A + c_1) - (m_1 x_B + c_1)$$
$$= m_1 x_A + c_1 - m_1 x_B - c_1 = m_1 x_A - m_1 x_B$$
$$= m_1 (x_A - x_B)$$

Hence, $m_1 = (y_A - y_B)/(x_A - x_B)$. Then $c_1 = y_A - m_1 x_A = y_B - m_1 x_B$. Thus, given the coordinates of any two points, we can calculate the parameters of the line that joins them (Box 3.2).

Box 3.2 — The Line Joining Two Points

Consider two points AB with coordinates

$$A\ (1234.56,\ 2345.67) \text{ and } B\ (1296.32,\ 2417.38)$$

Then
$$m_1 = (y_A - y_B)/(x_A - x_B)$$
$$= (2345.67 - 2417.38)/(1234.56 - 1296.32)$$
$$= (-71.71)/(-61.76) = +1.161108$$
$$c_1 = y_A - m_1 x_A = 2345.67 - 1.161108 * 1234.56$$

or
$$c_1 = y_B - m_1 x_B = 2417.38 - 1.161108 * 1296.32.$$

In both cases, this gives $c_1 = 912.21$ (giving an independent check). Hence, the line joining A to B is $y = 1.161108 x + 912.21$.

ALGEBRA — TREATING NUMBERS AS SYMBOLS

We can now calculate the point of intersection of two straight lines, as shown in Box 3.3.

Box 3.3 — The Point of Intersection of Two Lines

Following on from the example in Box 3.2, for two points CD with

$$C\ (1300.24,\ 2351.77)\ \text{and}\ D\ (1212.45,\ 2431.78)$$

we have $m_2 = (-80.01)/(87.79) = -0.9113794$, $c_2 = 3536.78$
or for the line CD $y = -0.9113794\ x + 3536.78$.

The point of intersection (P in Figure 3.2) between the lines AB and CD must satisfy both these conditions. Then

$$\text{For } AB:\quad y = 1.161108\ x + 912.21$$

$$\text{For } CD:\quad y = -0.9113794\ x + 3536.78$$

Hence, at the point P:

$$1.161108*x + 912.21 = -0.9113794*x + 3536.78$$

Add $0.9113794*x$ to both sides of the equation and take 912.21 away from both sides:

$$1.161108*x + 0.9113794*x + 912.21 - 912.21$$

$$= -0.9113794*x + 0.9113794*x + 3536.78 - 912.21$$

Or $2.072487*x = 2624.57$.

Dividing both sides by 2.072487, $x = 1266.39$.

Hence, $y = 1.161108*1266.39 + 912.21$ or $y = 2382.62$

The coordinates of P are therefore **(1266.39, 2382.62)**.

Thus, using the principles of arithmetic given at the start of Chapter 2, we have found the coordinates of the point of intersection of two lines.

The relationship $y = mx + c$ is a general expression for a straight line in rectangular Cartesian coordinates. 'y' is said to be the *dependent* variable and 'x' the *independent*, meaning that for any chosen value of 'x,' we can obtain a unique value of 'y.' In fact, we could have written the equation in the form $x = ny + d$ (where $n = 1/m$ and $d = -c/m$). 'x' would then be dependent on 'y.'

In Figure 3.4, let the line AB be $y = mx + c$. Consider the line $y = mx + c + 1$. For this line, for any value of x, the value of y is precisely 1 unit above the value in the line AB. This means that this new line is parallel to AB but 1 unit above it. More generally, every line $y = mx + d$, where d has any fixed value, will be parallel to AB. 'm' is a measure of the slope or angle and has the same value for all parallel lines.

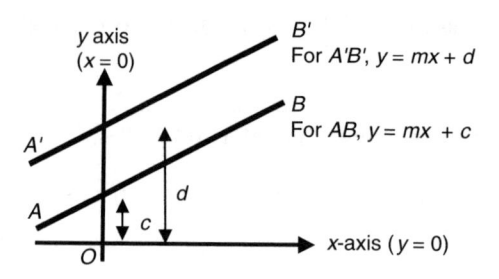

Figure 3.4 Parallel lines.

An example of an application for the equation $y = mx + c$ arises when drawing lines on a map that cross the map sheet edge (at P in Figure 3.5) or any square window drawn round an area — a process known as *clipping*. The lines that cross the boundary of the area being displayed must be clipped so that all sections of the original data that lie outside the area under consideration can be discarded.

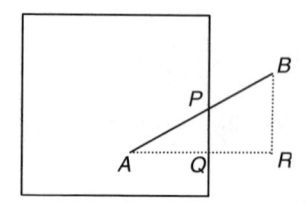

Figure 3.5 Clipping to a window.

Box 3.4 — Intersection at Map Sheet Edge (1)

Referring to Figure 3.5 and using the coordinates of A and B as before with $A(1234.56, 2345.67)$ and $B(1296.32, 2417.38)$. In Box 3.2 we showed that

$$y = 1.161108 * x + 912.21$$

Now, if, for example, the edge of the map sheet is on a grid line, QP whose x value $x_P = 1250$

then at that point $y_P = 1.161108 * 1250 + 912.21$

$$= 2363.60.$$

A computer-driven graphical plotter could be told to move from A (1234.56, 2345.67) to P (1250.00, 2363.60). It would then stop plotting at the map sheet edge, clipping the line AB at P.

One way to compute the sections of line to be included is given in Box 3.4. There is, however, an alternative approach to the truncation of the line AB that does not require the calculation of the equation $y = mx + c$. It is based entirely on the idea of *scale*. Consider the two triangles APQ and ABR in Figure 3.6.

ALGEBRA — TREATING NUMBERS AS SYMBOLS

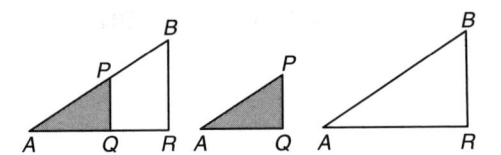

Figure 3.6 Similar triangles.

The two triangles APQ and ABR in Figure 3.6 are exactly the same shape. They only differ in size and are said to be similar. *Similar triangles* do not necessarily have one angle a right angle; the key point is that they have corresponding angles of the same size. In fact, they just differ in terms of scale. This means that all linear distances are in the same proportion or ratio. If the *scale factor* between them is some number '*s*' then

$$AP = s*AB, \quad AQ = s*AR \quad \text{and} \quad PQ = s*BR$$

Alternatively,

$$\frac{AP}{AB} = \frac{AQ}{AR} = \frac{PQ}{BR} = s$$

This will hold for similar triangles even when the triangle does not contain a right angle. In the case of Figure 3.5, we merely have to scale down BR to PQ to obtain the y value of P, as shown in Box 3.5.

Box 3.5 — Intersection at Map Sheet Edge (2)

Using the same numbers as in Box 3.4, if R has the same x value as B and y value as A then the distance

$$AR = x_B - x_A = 1296.32 - 1234.56 = 61.76$$
$$BR = y_B - y_A = 2417.38 - 2345.67 = 71.71$$
$$AQ = x_P - x_A = 1250.00 - 1234.56 = 15.44$$

Hence,

$$AQ/AR = 15.44/61.76 = 0.25 = s$$
$$QP = s*BR = 0.25*71.71 = 17.93$$

Or

$$PQ = y_P - y_A = y_P - 2345.67 = 17.93$$

Hence,

$$y_P = 2363.60$$

Once again, the coordinates of P are **(1250.00, 2363.60)**.

3.3 POINTS IN POLYGONS

If we express the equation of the line through the two points $A\ (x_A, y_A)$ and $B\ (x_B, y_B)$ in the form

$$(y - y_A) = \frac{(y_B - y_A)}{(x_B - x_A)} * (x - x_A)$$

then it is easy to test whether any point is above or below the line AB. To do so, take any point $C\ (x_C, y_C)$ and calculate

$$y = y_A + \frac{(y_B - y_A)}{(x_B - x_A)} * (x_C - x_A)$$

If the answer is less than y_C, then C is above AB and if it is greater, then C lies below the line from A to B. Simple tests like this are useful when determining whether two points are on the same or opposite sides of a line.

This also provides one way to test whether a point $P\ (x_P, y_P)$ lies within or outside a polygon. Consider the polygon $ABCDEF$ in Figure 3.7 where the coordinates of A are (x_A, y_A), of B are (x_B, y_B), etc. Firstly, one should check whether x_P lies between the maximum and minimum values of x for the whole polygon, which in this case means that $x_A \leq x_P \leq x_E$ and similarly $y_F \leq y_P \leq y_D$ in order to reduce the amount of computation. (The notation "$a \leq b$" means that 'a' is less than or equal to 'b.') Testing whether something is greater or less than a specified value is a relatively trivial computation.

Next, consider the section of the horizontal line through P that has x values greater than x_P. Let this be PQ where Q is such that $x_Q > x_E$ and $y_Q = y_P$. The number of occasions that this line crosses a side of the polygon must be counted.

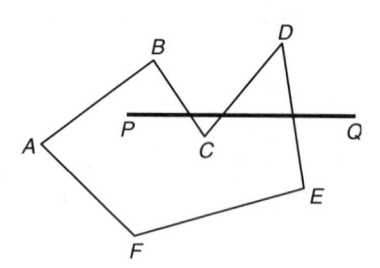

Figure 3.7 Point-in-polygon.

In Figure 3.7, a series of checks must be carried out on each of the lines AB, BC, CD, etc., that form the polygon. In the case of AB, y_P (the y value for the horizontal line from P to Q) is greater than y_A and less than y_B, so the line PQ will intersect AB. From the coordinates of A and B, we can work out the point on the line AB where the y value is y_P; the x value will be seen to be less than x_P and it can be ignored.

For the lines BC, CD, and DE, their intersection of PQ has an x-value greater than x_P and therefore they will count. Each of the lines EF and FA are wholly below P and can be ignored. At the end, if the number of lines that are crossed to the right of P is an even number, then P will be outside the polygon; if the number is odd (as in this case), the point P will be inside the polygon. Care must be taken if the line PQ goes exactly through one of the corner points as this can upset the count; this can, however, be avoided by a minor adjustment in the coordinates of P, for example, if the vertices are given to two decimal points (0.01) then by treating P to three decimal points (rounding it to 0.005) the problem will not arise.

3.4 THE EQUATION FOR A PLANE

We can extend the equation of a line to describe a plane. Consider the equation

$$z = mx + ny + c$$

For every fixed value of y (e.g., $y = d$),

$$z = mx + (nd + c) = mx + c'$$

where $c' = nd + c$ and is a constant. This is the equation for a straight line in the plane $y = d$. Similarly, for every fixed value of x, we have a set of straight lines. If z is regarded as the axis in the third dimension, then

$$z = mx + ny + c$$

represents a *plane* surface (Figure 3.8).

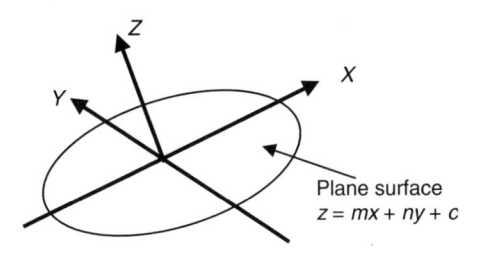

Figure 3.8 A plane surface.

3.5 FURTHER ALGEBRAIC EQUATIONS

The manipulation in Box 3.6 is an example of rearranging equations. Provided that we carry out the same operation on both sides of the equation, by adding, subtracting, multiplying, or dividing by the same amount, we shall not alter the basic relationship. This is a very important principle that helps solve many mathematical equations (see Box 3.7).

The only thing that we cannot do is divide both sides by zero, for although $2*0 = 1*0$, it does not follow that $2 = 1$.

Box 3.6 — The Intersection of Two Planes

When two planes intersect we have equations of the form

$$z = m_1 x + n_1 y + c_1, \quad z = m_2 x + n_2 y + c_2$$

Subtract one equation from the other:

$$z - z = m_1 x + n_1 y + c_1 - (m_2 x + n_2 y + c_2)$$

or

$$0 = m_1 x + n_1 y + c_1 - m_2 x - n_2 y - c_2$$

Add $n_2 y$ to both sides of the equation:

$$n_2 y = m_1 x + n_1 y + c_1 - m_2 x - n_2 y - c_2 + n_2 y$$

Subtract $n_1 y$ from each side of the equation:

$$n_2 y - n_1 y = m_1 x + n_1 y + c_1 - m_2 x - n_2 y - c_2 + n_2 y - n_1 y$$
$$= m_1 x + c_1 - m_2 x - c_2$$

or

$$(n_2 - n_1) y = (m_1 - m_2) x + (c_1 - c_2)$$

Divide both sides by $(n_2 - n_1)$

$$y = \left(\frac{(m_1 - m_2)}{(n_2 - n_1)} x + \frac{(c_1 - c_2)}{(n_2 - n_1)} \right)$$

Replace $((m_1 - m_2)/(n_2 - n_1))$ by 'm' and $+ ((c_1 - c_2)/(n_2 - n_1))$ by 'c.' The result is $y = mx + c$. This has already been shown to be the equation of a straight line. Thus, two planes intersect in a line.

Box 3.7 — Rule 9

Provided that you add, subtract, multiply, or divide an equation by the same amount on both sides of the equals sign, the relationship continues to hold.

All of the above relationships are examples of equations. An *equation* is a formula that asserts that two expressions have the same value. Thus, $y = mx + c$ is an *identical equation* or *identity* that is true for all values of the variables that lie along the line that is determined by the constants 'm' and 'c.' A *conditional equation* is one that is only true for certain values of the variables. For instance, "$x^2 - x = 2$" is a conditional equation that is only true when $x = 2$ or $x = -1$.

ALGEBRA — TREATING NUMBERS AS SYMBOLS

Two particular forms of equation commonly occur in geomatics — simultaneous equations and quadratic equations. A *simultaneous equation* is one of at least a pair of equations that must be satisfied simultaneously. An example was given in Box 3.3 where there were two straight lines that intersected. At the point of intersection P, the values of x and y had to satisfy the equation for the line AB and the equation for the line CD simultaneously. If there are two unknowns (here x and y), then there must be at least two independent equations in order that their points of intersection can be found and the equations can be solved. For three unknowns (x, y, z), there must be at least three independent equations, each of which must be satisfied. (In Chapter 11, we consider what happens when there are more than the minimum number of equations.)

A *quadratic* is a function in which the variables may be "raised to the power of two" or be of "second degree" — for example, x^2, y^2, or xy (but not xy^2 which is of the third degree). "$x^2 + y^2 = r^2$" is an example of a quadratic expression, one interpretation of which is that r is the radius of a circle with center $(0, 0)$ and (x, y) is any point on the circle. The relationship $x^2 + y^2 = r^2$ only exists in real terms when both x and y are equal to or greater than $-r$ and less than or equal to $+r$. This may be written in the form

$$-r \leq x \leq +r \quad \text{and} \quad -r \leq y \leq +r$$

or

$$|x| \leq r \quad \text{and} \quad |y| \leq r$$

where $|x|$ and $|y|$ mean the positive values of x and y. $|x|$ and $|y|$ are called the *modulus* of 'x' and modulus of 'y' or "mod x" and "mod y."

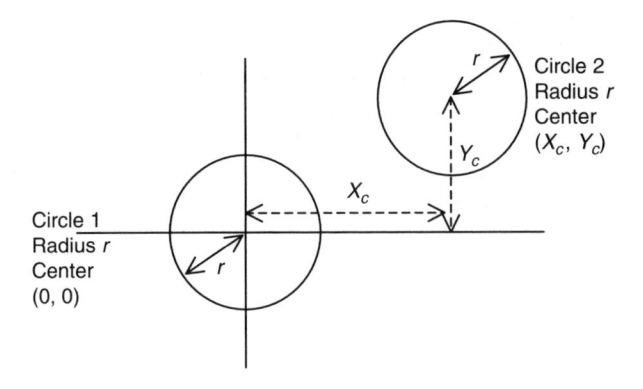

Figure 3.9 The equation for a circle.

For a circle of radius r around the point X_c, Y_c (Circle 2 in Figure 3.9) the general equation would take the form

$$(x-X_c)^2 + (y-Y_c)^2 = r^2$$

or

$$x^2 - 2X_c x + X_c^2 + y^2 - 2Y_c y + Y_c^2 = r^2$$

or

$$x^2 + y^2 + ax + by + c = 0$$

where a, b, and c are constants with

$$a = -2X_c, \quad b = -2Y_c, \quad \text{and} \quad c = X_c^2 + Y_c^2 - r^2$$

The term "quadratic" is commonly associated with "*quadratic equation.*" A quadratic equation takes the form

$$ax^2 + bx + c = 0$$

or

$$ax^2 + bx^1 + cx^0 = 0,$$

in which the power of x takes the integer values 0, 1, and 2. Because it is of second degree, there are two possible solutions. Mathematics solves such equations in a variety of ways. For example, the equations can be re-arranged by dividing every term by 'a' (assuming 'a' is not zero). Hence

$$x^2 + (b/a)x + (c/a) = 0$$

or

$$x^2 + (b/a)x = -(c/a)$$

Next, we add a number 'n' so that $x^2 + (b/a)x + n$ becomes a *perfect square*, that is, it is some value raised to the power 2. Let this value be $(x + p)$ so that we have to find 'p' such that $(x + p)^2 = x^2 + (b/a)x + n$. We do not of course know 'n' at this stage but $(x + p)^2 = x^2 + 2px + p^2$. Hence, we need to choose 'p' so that $2p = (b/a)$ and $p^2 = n$. This means that $p = (b/2a)$ and hence $n = p^2 = b^2/4a^2$. This gives a perfect square $(x + b/2a)^2$ or $x^2 + (b/a)x + b^2/4a^2$.

If we write our original equation $ax^2 + bx + c = 0$ as $x^2 + (b/a)x = -(c/a)$ and then add $b^2/4a^2$ to both sides, we obtain

$$x^2 + (b/a)x + b^2/4a^2 = b^2/4a^2 - (c/a)$$

or

$$(x + b/2a)^2 = b^2/4a^2 - (c/a) = b^2/4a^2 - (4ac/4a^2) = (b^2 - 4ac)/4a^2$$

Taking the square roots $(x + b/2a) = \pm \{\sqrt{(b^2 - 4ac)}\}/2a$, or

$$x = \{-b \pm \sqrt{(b^2 - 4ac)}\}/2a.$$

Note that if $b^2 < 4ac$, then $(b^2 - 4ac)$ is negative and since when using real numbers there is no such thing as the square root of a negative number, there is no solution to the problem since there is no real value of x that satisfies $ax^2 + bx + c = 0$ (Box 3.8). When solving simultaneous equations, a similar approach is used in which we alter both sides of an equation (Box 3.9).

Box 3.8 — Solving a Quadratic

As shown in the main text, if $ax^2 + bx + c = 0$,

$$x = \{-b \pm \sqrt{(b^2 - 4ac)}\}/2a$$

Consider $3x^2 + 4x + 1 = 0$. Here, $a = 3$, $b = 4$, $c = 1$.
Hence $x = \{-4 \pm \sqrt{(16 - 12)}\}/6$.
So $x = (-4 + 2)/6 = -1/3$, or $x = (-4 - 2)/6 = -1$.
The two solutions to $3x^2 + 4x + 1 = 0$ and $x = -1/3$ or $x = -1$.

Consider two lines $ax + by + c = 0$ and $dx + ey + f = 0$. Multiply both sides of the first equation by 'd' and both sides of the second equation by a. This gives

$$adx + bdy + cd = 0,$$
$$adx + aey + af = 0$$

Subtracting $adx + bdy - adx - aey + cd - af = 0$

Or $y(bd - ae) = (af - cd)$. Or $y = - (af - cd)/(ae - bd)$.

Similarly $x = (bf - ce)/(ae - bd)$ or

$$\frac{x}{(bf - ce)} = \frac{-y}{(af - cd)} = \frac{1}{(ae - bd)}$$

Box 3.9 — Intersecting Lines by Simultaneous Equations

As shown in the main text, if there are two equations

$$ax + by + c = 0 \text{ and } dx + ey + f = 0$$

Then

$$\frac{x}{(bf - ce)} = \frac{-y}{(af - cd)} = \frac{1}{(ae - bd)}$$

Consider the two lines $3x + 4y - 10 = 0$ and $5x - 2y - 8 = 0$. These will intersect, where $x/(-52) = -y/(26) = 1/(-26)$ so $x=2$ and $y=1$. The values $x=2$ and $y=1$ satisfy both equations. Hence, they must represent the point where the two lines intersect.

3.6 FUNCTIONS AND GRAPHS

A *function* is a relationship in which given the value of one or more variables, the overall value of the function can be calculated. For example, $y = f(x)$ can be read as "y is a function of x" with y as the value of the function and x the *argument*. The function can take any mathematical form such as

$$y = ax + b \quad \text{(which is a \emph{line})}$$
$$y = ax^2 + bx + c \quad \text{(a \emph{quadratic} or second degree curve)}$$
$$y = ax^3 + bx^2 + cx + d \quad \text{(a \emph{cubic} or third degree curve) etc.}$$

A particular group of functions are called *polynomials*. A polynomial is essentially an expression that contains two or more terms such as

$$x^3 + 3x^2y + 3xy^2 + y^3 + x^2 + 3y^2 + xy + x + y + 4$$

More specifically, the term is used to describe a relationship such as

$$y = a + bx + cx^2 + dx^3 + ex^4 + fx^5 + \cdots$$

where a, b, c, d, e, f, etc., have fixed values. There may be a finite number of terms, for example, where

$$y = (1+x)^4 = 1 + 4x + 6x^2 + 4x^3 + x^4$$

Alternatively, there may be an infinite number of terms in which case it will either generate an infinite number or else converge onto a specific value.

The term *convergence* means that however many terms are added, the function will simply grow closer and closer to a specific value.

Thus, $e = 1 + 1/1! + 1/2! + 1/3! + \cdots$ has an infinite number of terms but it never becomes larger than a certain value, while the series of terms $1 + 2 + 3 + \cdots$ just becomes larger and larger and thus is not convergent.

A *series* is a sequence of terms that may be finite or infinite. An example of a series is $S = a_0 + a_1 + a_2 + a_3 + \cdots + a_{(n-1)} + a_n$ where n is a positive integer. It is often written as

$$\Sigma a_i \text{ or } \sum_{i=0}^{n} a_i$$

where the symbol Σ (the Greek letter Sigma) means "the sum of." In practice, an infinite series will only have a finite sum if the series ($a_0, a_0 + a_1, a_0 + a_1 + a_2, \ldots$, etc.) converges.

Functions may involve one dependent and one independent variable or may have several independent variables such as: $z = f(x, y)$, meaning that z is a function of x and y where x and y are both independent variables. If z is dependent on x and y then we have a surface. If z is linearly dependent on x and y (i.e., $z = f(x,y) = ax^1 + by^1 + c$) then, as we have seen above, this is a plane surface. If $z = ax^2 + by^2 + cx + dy + e$, then we have a second degree or quadratic surface.

ALGEBRA — TREATING NUMBERS AS SYMBOLS

If $y = f(x)$ so that y is a function of x, then there is a unique value for y for every value of x. If the relationship between y and x is such that it is also possible to determine x uniquely given the value of y then the relationship is said to have an *inverse function*, written as f^{-1}. Thus, if $y = f(x) = ax + b$, then $x = f^{-1}(y) = (y-b)/a$.

This is often not possible. For instance, if $y = x^2$ then $x = +\sqrt{y}$ or $x = -\sqrt{y}$ and there are two possible relationships. Hence, there is no inverse function. In general, functions are mathematical relationships between two (or more) variables that may be in the form of one-to-one or one-to-many or many-to-many. We have seen that the equation for a circle centred at the origin is $x^2 + y^2 = r^2$ where r = radius. For any value of x that represents a point on the circle, there are two possible values of y (one positive and one negative); similarly for any value of y, there are two values of x and thus there is a one-to-two relationship between x and y and between y and x.

Relationships between two variables are often better expressed graphically than numerically since many people find a visual image more easy to understand than a mathematical equation. A *graph* is a drawing showing the relationship between certain sets of quantities by means of points or lines plotted with respect to a set of coordinate axes. As an example of a graph, consider the function $y = 0.5x^2$.

The values of points on the graph can be calculated by choosing a series of values for 'x' and computing 'y.' Every pair of (x, y) coordinates can then be plotted. Intermediate points can be calculated as necessary (Table 3.1).

Table 3.1 Values of y and x for $y = 0.5x^2$

x	0	±0.2	±0.4	±0.6	±0.8	±1	±1.2	±1.4	±1.6	±1.8	±2
y	0	0.02	0.08	0.18	0.32	0.5	0.72	0.98	1.28	1.62	2

When plotted, the coordinates form the curve as shown in Figure 3.10. In general, when drawing a graph, a series of points are plotted and are then joined up by straight lines. A circle may be drawn on a piece of paper using a compass, resulting in a nice smooth shape; but in most cases this is not possible and as described in Chapter 1, curves are either plotted in the form of dots (the raster method) or as a series of short vectors with the precise location of points on the curve being calculated from the

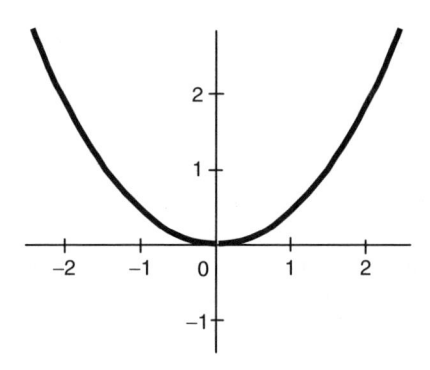

Figure 3.10 Graph of the function $y = 0.5x^2$.

algebraic formulae. In practical terms, not all points can be plotted as they may extend far beyond the limits of any piece of paper. In Figure 3.10, if $x = 100$ then $y = 5000$; so the axes would have to be scaled down to accommodate such a range.

Figure 3.11 shows the curve for $y = 1/x$. When $x = 1$ then $y = 1$; when $x = 0.5$ then $y = 2$; when $x = 0.1$ then $y = 10$; when $x = 0.001$ then $y = 1000$; and so on. As 'x' becomes smaller and smaller and approaches zero from the positive side, 'y' goes off to plus infinity ($+\infty$). As 'x' approaches zero from the negative side 'y' goes off to minus infinity ($-\infty$).

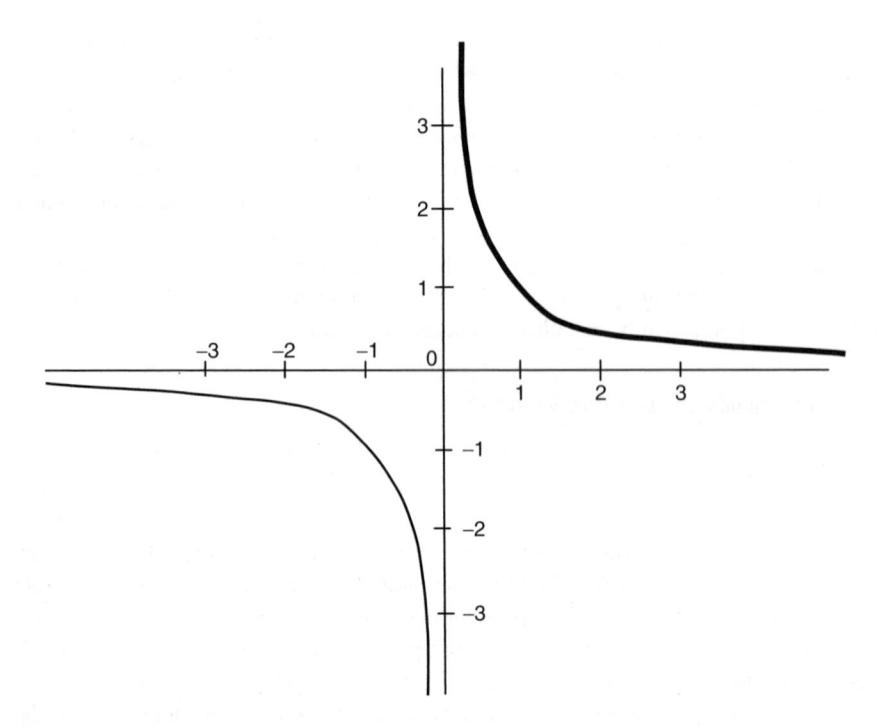

Figure 3.11 Graph of $y = 1/x$.

The value of y when $x = 0$ can only be expressed mathematically as $y = \infty$. It cannot be shown on any graph. It also has the peculiar property that it flips from $+\infty$ to $-\infty$ or *vice versa* as the value of x passes through zero.

3.7 INTERPOLATING INTERMEDIATE VALUES

Both the graphical and numerical approaches can be used to determine values in between or beyond those that have been measured. The process of determining intermediate values is known as *interpolation*. Taking a guess at what happens beyond what is already known is called *extrapolation*; for instance, we know approximately the number of people who were in the world at the end of each decade over the last

ALGEBRA — TREATING NUMBERS AS SYMBOLS

century and by plotting the numbers on a graph, it is possible to predict how many people will be in the world in AD 2020 or even AD 2100 assuming that the present trends continue. For the present, we shall only consider simple linear interpolation and delay until later any discussion on more complex forms.

In Figure 3.12, P represents a point interpolated along the straight line between A and B; Q is an extrapolation and assumes that AB continues in a straight line.

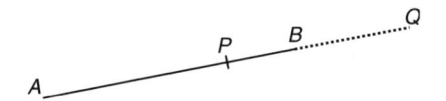

Figure 3.12 Linear interpolation and extrapolation.

The coordinates of intermediate points can be calculated on the basis of ratios. Thus, if in Figure 3.12 PB is one quarter of AB, then it divides AB in the ratio 3 to 1 so, by interpolation:

$$X_P = (3*X_B + X_A)/4 \quad \text{and} \quad Y_P = (3*Y_B + Y_A)/4$$

If Q is the point on the line AB such that $BQ = (1/2) BA$ then B divides AQ in the ratio 2 to 1 so $B = (2Q + A)/3$ or $Q = (3B - A)/2$. Thus,

$$X_Q = (3*X_B - X_A)/2 \quad \text{and} \quad Y_Q = (3*Y_B - Y_A)/2$$

We can use these types of relationship to show that in a triangle ABC the lines joining each apex to the midpoint of the opposite side all pass through one point (Figure 3.13). See Box 3.10.

Box 3.10 — Example of Interpolation. The Co-ordinates of the Centroid of a Triangle

In Figure 3.13 let D be the midpoint of BC. G divides AD in the ratio 2:1. (So that $GD = 1/3\text{rd of } AD$) Let the coordinates be A (x_A, y_A), B (x_B, y_B), and C (x_C, y_C). By proportion, the coordinates of D are

$$(x_D, y_D) = \left\{ \frac{x_B + x_C}{2}, \frac{y_B + y_C}{2} \right\}$$

The coordinates of

$$G = \left\{ \frac{(2x_D + x_A)}{3}, \frac{(2y_D + y_A)}{3} \right\} = \left\{ \frac{x_A + x_B + x_C}{3}, \frac{y_A + y_B + y_C}{3} \right\}$$

Note: Since this is symmetrical, it would be the same for BE where E is the midpoint of AC and CF where F is the midpoint of AB. The lines joining the vertices of a triangle with the midpoints of the opposite sides pass through one point G, known as the *centroid*.

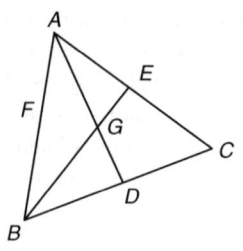

Figure 3.13 The midpoints of the sides of a triangle.

The process of reducing graphs to straight line sections helps in interpolation making the calculation of intermediate values relatively simple. An example of such a calculation was given previously in Figure 3.5 and the cutting off of a line at a map sheet edge or window boundary. Given two points, each one on either side of the required value, the coordinates of the intermediate point can be determined. On the other hand, if as in Figure 3.14, A is the top of a slope and B the bottom, linear interpolation will overestimate lower heights and underestimate the higher ones.

Figure 3.14 Interpolation of heights down a slope.

Intermediate values are often interpolated on the assumption that change is linear. In particular, this can happen when constructing lines of equal value such as contour lines. Nearby points of known value (such as spot heights) are joined up and the positions where a contour line crosses the lines of the resulting triangles are then interpolated. These positions in turn are joined up to form the contour. The procedure can be applied to any third-dimensional value such as air pressure or temperature at a point; but here we shall consider heights.

In Figure 3.15(a), the five spot heights have been joined by straight lines to form triangles. In Figure 3.15(b) the values equivalent to 25 m have been linearly interpolated along each line. Thus, between the 27.5 and 21.2 m spot heights, the level drops by 6.3 m. To drop to 25 m from 27.5 means to drop 2.5 m out of the total of 6.3 m. The contour crosses the line at a distance of (2.5/6.3) of its length. If the end points are (x_A, y_A) and (x_B, y_B), then the coordinates of the interpolated point will be

$$\left\{ \frac{(3.8x_A + 2.5x_B)}{6.3}, \frac{(3.8y_A + 2.5y_B)}{6.3} \right\}$$

The interpolated points can then be joined up either (c) by straight lines or (d) by a curve drawn freehand or by using a computer algorithm (Figure 3.15(c) and (d)).

ALGEBRA — TREATING NUMBERS AS SYMBOLS

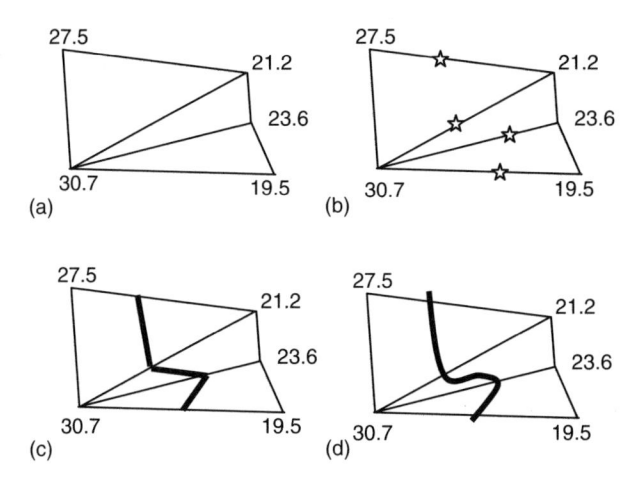

Figure 3.15 Interpolating contours between spot heights.

The assumption behind the interpolation is that the lines chosen are of a uniform slope. In fact, other triangles could have been formed as in Figure 3.16(a) and (b). Each will give a different contour line. Which is right?

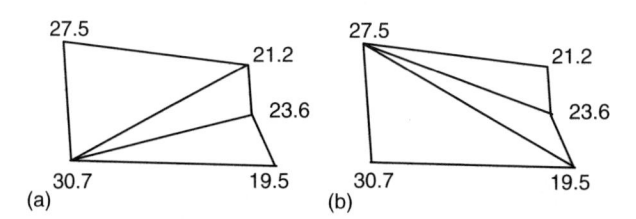

Figure 3.16 Alternative triangulation networks.

The answer is that neither will give a perfect answer, which is why contours are generally regarded as only being reliable within a band representing about a third of the contour interval (for instance, within just over 3 m for a 10 m contour interval). The main problem is with the assumption that height changes evenly between spot heights. In some cases, the surveyor may have made sure that points are chosen to fulfill this criterion but, in general, slopes are not uniform and a linear interpolation is just a convenient guess.

In Chapter 4, we will describe a technique that uses what is known as the Theissen Polygon to select good-shaped triangles. For the present, it should just be noted that the accuracy of the interpolation technique depends on (i) the accuracy of the original data; (ii) assumptions behind the interpolation; and (iii) the guesswork that enters into smoothing the contour line into an acceptable curve. We will explore ways in which a smooth curve can be fitted through a series of points in Chapter 8.

CHAPTER **4**

The Geometry of Common Shapes

CONTENTS

4.1   Triangles and Circles 43
4.2   Areas of Triangles 46
4.3   Centers of a Triangle 49
4.4   Polygons 51
4.5   The Sphere and the Ellipse 53
4.6   Sections of a Cone 55

4.1 TRIANGLES AND CIRCLES

Geometry is the study of constructible shapes. In this chapter, we shall review some of the shapes that occur in geomatics and occupy ordinary two- or three-dimensional (2D or 3D) "Euclidean" space. Euclid was a Greek mathematician of the 3rd century BC who worked out a series of axioms or postulations concerning points, lines, angles, surfaces, and volumes. From these, he derived 465 theorems.

His basic axioms included such statements as that for any two distinct points there is only one straight line that passes through them, and if three distinct points are not on a straight line, then there is only one plane that will pass through them. Euclid identified ten axioms but subsequently a further one was added, namely that only one straight line can be drawn parallel to a given line through any point not on that line.

In Euclidean space, the shortest distance between two points is a straight line. Two lines that are either parallel or intersect form a plane — in fact, parallel lines may be said to intersect at a point at infinity, an important consideration when drawing images of 3D objects in perspective on a plane (2D) surface, as discussed in Chapter 9.

The triangle is the simplest shape that is made up of straight lines. In fact, all 2D shapes can be regarded as being made up from a series of triangles, just as every curve can be thought of as a series of short straight lines. Although this can give rise to a number of errors, for example, when calculating an area enclosed by a curved line, the approximation can be adequate for many practical purposes.

Triangles come in all kinds of shapes and sizes but the basic fact is that the angles of a plane triangle add up to half a complete turn or 180°; the angles of a spherical triangle add up to more than 180°.

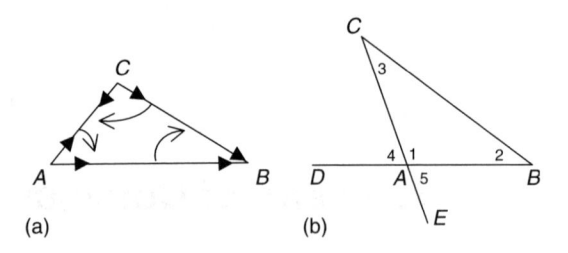

Figure 4.1 The angles of a triangle.

In the triangle ABC (see Figure 4.1(a)), at A face C then turn clockwise to B; at B turn clockwise to face backwards at C; at C turn clockwise to face A; you will have turned through the three angles ABC and half a complete turn. It is sometimes convenient to measure this in units of radians where π radians ('pi' radians) = 180°. Measures in radians are particularly important when using differential and integral calculus as discussed in Chapter 6. For the present, all angles will be considered as being in degrees, each degree being divided into 60 minutes (60′) and each minute containing 60 seconds (60″).

It is conventional for the angles of the triangle formed by the points A, B, and C to be designated by capital letters while the sides opposite the angles have lowercase letters a, b, and c. In Figure 4.1(b), the internal angles 2 and 3 are both less than 90° and are *acute* while the internal angle A is greater than 90° and is *obtuse*.

Note that since the angles $\angle 1 + \angle 2 + \angle 3 = 180°$ and also the angle marked '4' is such that $\angle 4 + \angle 1 = 180°$, then the external angle of the triangle is equal to the sum of the two internal opposite angles. The symbol '$\angle$' means "angle."

This also applies to the angle marked 5 in Figure 4.1(b). Thus, $\angle 4 = \angle 2 + \angle 3 = \angle 5$. In fact, whenever two lines intersect, the opposite angles are equal (Box 4.1).

Box 4.1 — Angles of a Triangle

Rule 4.1. The external angles of a triangle obtained by extending each of the sides equal the sum of their interior opposite angles

Rule 4.2. When two straight lines intersect, opposite angles are equal

---

Triangles with unequal sides are said to be *scalene*, while those that have two angles equal are called *isosceles*. In Figure 4.2, in triangle OAC if OP is perpendicular to AC and angle ACO = angle OAC, then it follows that the two triangles OPA and OPC have the same angles and a common side OP. They must therefore be exactly the same size and so $OC = OA$. An isosceles triangle such as OAC has two angles equal and also two sides equal. If all three sides are equal so that each angle = 60° then the triangle is *equilateral*.

Two triangles that have the same angles but are of different size are said to be *similar*. Similar triangles differ merely in scale. Triangles that are exactly the same size and shape are said to be *congruent*.

THE GEOMETRY OF COMMON SHAPES

Extending Rule 4.1, consider Figure 4.2 where the circle with center O passes through the points A, B, and C. Extend CO to D so that CD is a diameter of the circle. Also draw OQ perpendicular to AC at P. Any straight line that crosses a circle is called a *chord* (e.g., AC and CB in Figure 4.2), while the curved section that a chord cuts off is an *arc* (e.g., the curved sections AQC or ADB).

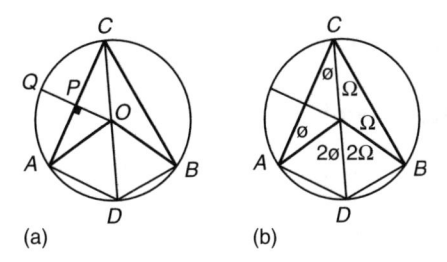

Figure 4.2 Angles subtended by arcs.

In Figure 4.2(a), $OA = OB = OC$ = the radius of the circle. Hence, triangles AOC and BOC are isosceles and $\angle OAC = \angle OCA$. Let this angle be ø ("phi"). Hence, $\angle AOD = 2ø$. Also, $\angle OCB = \angle OBC = \Omega$, say ("omega") and $\angle BOD = 2\Omega$. Thus, $\angle ACB = ø + \Omega$ while $\angle AOB = 2(ø + \Omega)$. This is true wherever C is along the section of the circle $AQCB$ (Box 4.2).

Box 4.2 — Angles Subtended by an Arc

Rule 4.3. The angle subtended by an arc at the center of a circle equals twice the angle subtended at the circumference

Rule 4.4. The angles subtended by an arc in the same segment of the circumference of a circle are all equal

---

In Figure 4.2(b), the external angle $AOB = 360 - (2ø + 2\Omega)$ will be twice the angle ADB, which therefore equals $180 - (ø + \Omega)$. Thus, $\angle ACB + \angle ADB = 180°$ as long as C and D are in opposite segments (a *segment* being a portion of a circle bounded by an arc and a chord; any chord divides a circle into two segments). It also follows that as in Figure 4.2 where the line CD is a diameter, the angle at the center $= 180°$ and hence the angle at the circumference would be $90°$. Thus, $\angle DAC$ and $\angle DBC$ both equal $90°$ (Box 4.3).

Box 4.3 — Angles in a Circle

Rule 4.5. The opposite angles of a quadrilateral bounded by a circle add up to $180°$

Rule 4.6. The angle subtended by a diameter at the circumference of a circle equals $90°$

Finally, in Figure 4.2 in triangles *OPA* and *OPC*, the angle at $P = 90°$ by construction (*OQ* was defined as being perpendicular to *AC*). Angle *OAP* = angle *OCP*, since triangle *AOC* is isosceles, *OP* is a common side, and $AO = CO$ are radii of the circle. Hence, triangle *OAP* is congruent to triangle *OCP* and thus $AP = PC$. In other words, *O* lies on the perpendicular bisector of the chord. Since this applies to each of the sides of a triangle, it follows that the circle that passes through each apex of a triangle has as its center the point where the three perpendicular bisectors of the sides of the triangle meet (Box 4.4).

Box 4.4 — Bisecting a Chord

Rule 4.7. The perpendicular bisector of a chord passes through the center of the circle

---

4.2 AREAS OF TRIANGLES

We shall return to aspects of the geometry of the circle shortly but for the present consider the triangle *ABC* in Figure 4.3. The triangle is half a rectangle, whose area is its base times its height. Therefore, the area of a triangle = $(1/2)*base*height = 0.5hc$.

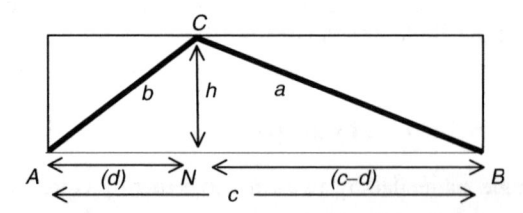

Figure 4.3 The area of a triangle.

We can extend the calculations of the area of a triangle to the *trapezium*, which is a four-sided figure with two sides parallel as shown in Figure 4.4. In Figure 4.4, the trapezium is made up of two triangles (*ABP* and *QCD*) and a rectangle PBCQ (Box 4.6). Let $BP = CQ = h$, $AP = t$, $PQ = u$, and $QD = v$. Then

$$\text{area} = (1/2)\, t*h + u*h + (1/2)\, v*h$$

$$= (1/2)\, (t + u + v)*h + (1/2)\, u*h = (1/2)\, (AD + BC)\, h$$

= half the sum of the parallel sides times the distance between them

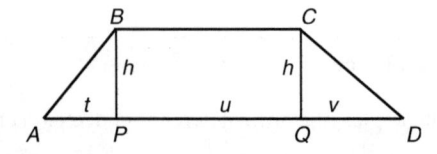

Figure 4.4 The area of a trapezium.

Box 4.5 — The Area of a Triangle from its Semiperimeter

The triangle ABC in Figure 4.3 may be divided by the perpendicular CN into two areas with $AN = d$ and $NB = (c-d)$.

The area of triangle ABC = (1/2) base*height = $(1/2)hc$.

Using Pythagoras for triangle ANC $\quad b^2 = h^2 + d^2$ or $h^2 = b^2 - d^2$.

Also, in triangle BNC, $\quad a^2 = h^2 + (c - d)^2 = h^2 + c^2 - 2cd + d^2$

$$= h^2 + d^2 + c^2 - 2cd.$$

Since $b^2 = h^2 + d^2$ this will equal $b^2 + c^2 - 2cd$.

Rearranging, $\quad 2cd = b^2 + c^2 - a^2$,

or $\quad d = (b^2 + c^2 - a^2)/2c$.

Hence,

$$h^2 = b^2 - d^2 = (b - d)(b + d) \text{ (the difference of two squares)}$$

$$= (b - (b^2 + c^2 - a^2)/2c)(b + (b^2 + c^2 - a^2)/2c)$$

$$= \{(2bc - b^2 - c^2 + a^2)/2c\} \{(2bc + b^2 + c^2 - a^2)/2c\}$$

$$= \{a^2 - (b - c)^2\} \{(b + c)^2 - a^2\}/4c^2$$

These again are the differences of two squares. Hence

$$h^2 = \{(a - b + c)(a + b - c)(b + c - a)(a + b + c)\}/4c^2, \text{ or}$$

$$4c^2h^2 = \{(a + b + c - 2b)*(a + b + c - 2c)*(a + b + c - 2a)$$

$$*(a + b + c)\}$$

$$= (2s - 2b)(2s - 2c)(2s - 2a)(2s) \quad \text{where } 2s = a + b + c$$

$$= 16s(s-a)(s-b)(s-c)$$

Or, dividing both sides by 16,

$$(1/4)\, c^2h^2 = s(s-a)(s-b)(s-c)$$

Taking the square roots

$$(1/2)ch = \text{the area of } ABC = \sqrt{s(s - a)(s - b)(s - c)}$$

where s is the semiperimeter of the triangle. Thus, the area of a triangle = $\sqrt{s(s - a)(s - b)(s - c)}$ where 's' is half its perimeter

Box 4.6 — The Area of a Triangle and a Trapezium

Rule 4.8. The area of a triangle is "half base times height." It also equals $\sqrt{s(s-a)(s-b)(s-c)}$ where 'a', 'b', and 'c' are the lengths of the sides, and 's' is the semiperimeter $= (1/2)(a + b + c)$

Rule 4.9. The area of a trapezium equals half the sum of the length of the parallel sides times the distance between them (Box 4.5)

We can extend this simple relationship to the coordinates of A, B, and C in Figure 4.5, where the area of triangle ABC = trapeziums $\{A'ACC' + C'CBB' - A'ABB'\}$. For the trapezium $A'ACC'$, $A'A = y$ coordinate of A, $C'C = y$ coordinate of C, while the distance between is $x_C - x_A$. The area $= \frac{1}{2}\{(y_A + y_C)(x_C - x_A)\}$. Likewise, for trapezium $C'CBB'$, the area $= \frac{1}{2}\{(y_C + y_B)(x_B - x_C)\}$, and for $A'ABB'$ the area $= \frac{1}{2}\{(y_A + y_B)(x_B - x_A)\}$. Combining the above and simplifying, the area of the triangle ABC is $\frac{1}{2}\{x_A y_B + x_B y_C + x_C y_A - y_A x_B - y_B x_C - y_C x_A\}$.

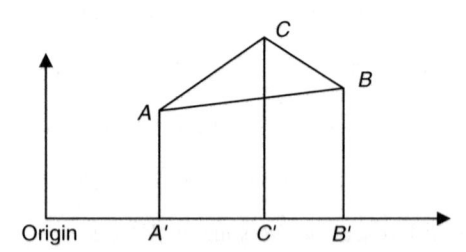

Figure 4.5 The area of a triangle by co-ordinates.

In fact, since every polygon can be subdivided into a series of triangles, this sequence can be extended. Thus, for a six-sided polygon with points A, B, C, D, E, and F, the area would be

$$\frac{1}{2}\{x_A y_B + x_B y_C + x_C y_D + x_D y_E + x_E y_F + x_F y_A$$

$$- y_A x_B - y_B x_C - y_C x_D - y_D x_E - y_E x_F - y_F x_A\}$$

For an eight-sided polygon A, B, C, D, E, F, G, and H, the area would be (Box 4.7)

$$\frac{1}{2}\{x_A y_B + x_B y_C + x_C y_D + x_D y_E + x_E y_F + x_F y_G + x_G y_H + x_H y_A$$

$$- y_A x_B - y_B x_C - y_C x_D - y_D x_E - y_E x_F - y_F x_G - y_G x_H - y_H x_A\}$$

and so on.

THE GEOMETRY OF COMMON SHAPES   49

> **Box 4.7 — The Area of a Polygon**
>
> **Rule 4.10.** To obtain the area of a polygon with corners (x_A, y_A), etc., multiply the x value of each point by the y value of the next point and then subtract the y value of each point times the x value of the next point to give twice the area.
> As an example, for an eight-sided figure, the area is
>
> $$\tfrac{1}{2}\{x_A y_B + x_B y_C + x_C y_D + x_D y_E + x_E y_F + x_F y_G + x_G y_H + x_H y_A$$
> $$- y_A x_B - y_B x_C - y_C x_D - y_D x_E - y_E x_F - y_F x_G - y_G x_H - y_H x_A\}$$

4.3 CENTERS OF A TRIANGLE

In Chapter 3, Box 3.10, we showed that the lines joining the corners of a triangle to the middle of their opposite sides pass through a point G known as the *centroid*. On the basis that the area of a triangle is half its base times its height, it follows that for the left-hand area in Figure 4.6, area of ACF = area of FCB and that in effect the triangle ABC is balanced about the line CF, similarly for the lines from A and B. Thus, G is the point of balance or center of gravity of the triangle. Its coordinates were shown in Box 3.10 to be $\{\tfrac{1}{3}(x_A + x_B + x_C), \tfrac{1}{3}(y_A + y_B + y_C)\}$.

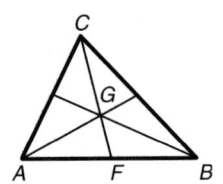

Figure 4.6 The centroid.

In Figure 4.7, AD is constructed so as to be perpendicular to BC, and BE to be perpendicular to AC. The two lines intersect at O and line CO meets AB at H.

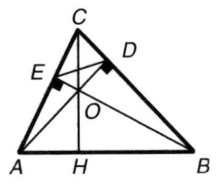

Figure 4.7 The orthocenter.

Since $\angle AEB = 90° = \angle ADB$, then a circle with AB as diameter passes through E and D (Rule 4.6); hence, $\angle EDO = \angle EBA$ (angles in the same segment — Rule

4.4) $= 90 - A$. Since $\angle ODC = 90°$ by construction, $\angle EDC = A$. But E and D also lie on a circle with diameter OC; hence, $\angle EDC = \angle EOC = A$. In triangle EOC, $\angle OEC = 90°$; hence, $\angle OCE = 90 - A$. Now in triangle CAH, $\angle CAH = A$ and $\angle ACH = \angle OCE = 90 - A$; hence, $\angle AHC = 90°$. Thus, CH is perpendicular to AB and the perpendiculars from each point of a triangle to its opposite side intersect at one point, known as the *orthocenter*.

In Figure 4.8, let the line CI be the bisector of the angle ACB and let AI bisect CAB. Let P, Q, and R be the feet of the perpendiculars from I to the corresponding sides. Then in triangles IPC and IQC, $\angle ICP = \angle ICQ$ since IC bisects PCQ, and $\angle IPC = 90° = \angle IQC$. Thus, the two triangles have the same angles and a common side (CI) and must therefore be congruent. This means that $IP = IQ$. Similarly, $IP = IR$ and triangles IRB and IQB are congruent so that IB is the bisector of angle ABC.

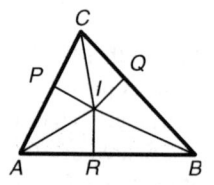

Figure 4.8. The incenter.

I is the *incenter* of the triangle and is the point where the bisectors of all three angles meet. A circle can be drawn around this point, which touches each of the three sides of the triangle (Figure 4.10(a)). Finally, the perpendicular bisectors of each side also meet in a point known as the *circumcenter* (CC in Figure 4.9 and Figure 4.10(b)). It is the center from which a circle can be drawn through all three vertices.

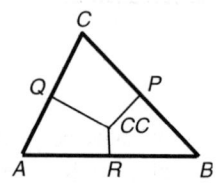

Figure 4.9 The circumcenter.

(a)   (b)

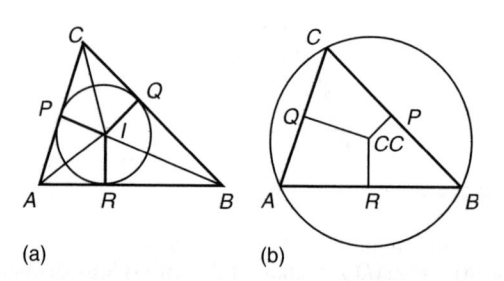

Figure 4.10 Inscribed and circumscribed circles.

THE GEOMETRY OF COMMON SHAPES

Circles have a number of well-known properties, such as the length of the circumference $= 2\pi r$ where $r =$ length of the radius. The area $= \pi r^2$. A line that just touches the circle (e.g., AB at point R in Figure 4.10(a)) is called a *tangent*. Any line other than a tangent will either not touch the circle at all or else will intersect it at two points. A tangent and the radius to the point where it touches the circle are at 90° and this radius is said to be *normal* to the tangent (Box 4.8).

Box 4.8 — Centers to a Triangle

1. The perpendiculars from the vertices to the opposite sides all pass through the orthocenter.
2. The bisectors of each angle pass through the incenter.
3. The perpendicular bisectors of each side pass through the circumcenter.
4. The lines joining each vertex to the midpoint of the opposite side all pass through the centroid.

4.4 POLYGONS

A *polygon* is a closed figure bounded by three or more straight lines that do not cross and have the same number of vertices as there are sides. The polygon is said to be concave if any of its interior angles are greater than 180°; otherwise, it is said to be convex.

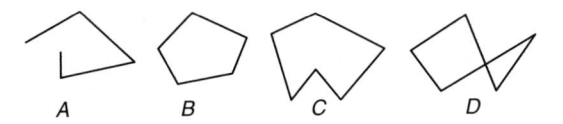

Figure 4.11 Straight-line figures.

In Figure 4.11, A is not a polygon, B is a convex polygon, C is a concave polygon, and D is essentially two polygons. The best-known examples of polygons are triangles, quadrilaterals (especially squares and rectangles), pentagons, and hexagons (six-sided figures). A contour line that closes back on itself and is constructed out of straight-line sections would also be a polygon. Any polygon can be broken down into a series of triangles, although even with a rectangle these are not unique (see Figure 4.12).

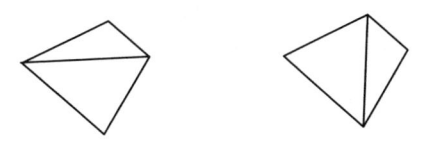

Figure 4.12 Two ways to divide a quadrilateral.

The three sides of the first triangle create internal angles that add up to 180°. A further point adds a further triangle and a further 180° to bring the sum of the internal angles to 360°. A five-sided polygon will have internal angles adding up to 540°, etc. Since the whole polygon can be thought of as a series of adjacent triangles, the sum of its interior angles will be $= (n-2)*180°$, where $n =$ the number of sides (Box 4.9).

> **Box 4.9 — The Angles of a Polygon**
>
> **Rule 4.11.** The sum of the interior angles of a polygon is $(n-2)*180°$ where $n =$ the number of sides.

Given any set of points, there are many ways in which they can be joined to form a set of triangles in a process known as *triangulation*. Until recently, the whole basis of map-making was constructed around a set of points between which the angles had been observed. The result was a network of polygons and triangles, some overlapping.

By measuring all three angles in each triangle, the shape of the mesh could be rigidly determined and there would be independent checks on the accuracy of the measurements since for each triangle the measured angles had to add up to 180°. The overall scale was determined by measuring at least one of the sides of the network (Figure 4.13).

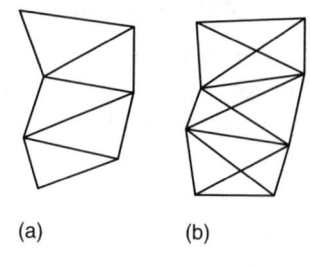

Figure 4.13 Triangulation networks.

Although the first triangulation networks were observed during the 18th century, it was during the latter part of the 19th and most of the 20th centuries that many areas of the globe became covered by such networks. The fact that the surface of the Earth is curved meant that none of the networks of triangles were flat and hence some corrections had to be applied to the measured angles and distances in order to calculate the coordinates of the triangulation or "trig" stations. The relationships between angles and distances on a curved surface and their corresponding values on the flat are discussed further in Chapters 5 and 9.

Triangulation is also used when interpolating heights from a series of irregular points in a process known as triangulated irregular network (TIN). TIN has become the generic term that is used in GIS to describe a way in which intermediate height values can be interpolated. Although height is the most commonly used third dimension in geomatics, the technique can be extended to other characteristics at a point, such as air pressure or the strength of the magnetic field. The process involves the conversion of a set of irregularly distributed points of known third-dimensional value ("spot heights") into a network of triangles within which further height values can be interpolated, as discussed in Chapter 3 (see Figures 3.15 and 3.16).

THE GEOMETRY OF COMMON SHAPES

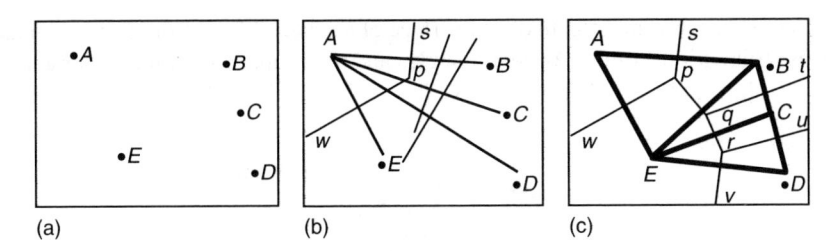

Figure 4.14 Theissen polygons and the Delaunay triangles.

Consider a series of points A, B, C, D, E in Figure 4.14(a) for which the eastings, northings, and height or (x, y, z) values are known. The points can be joined by straight lines and the intermediate heights can be interpolated. The question here is "which points should be connected?" A common method is known as the Delaunay triangulation. It is related to the creation of a series of polygons known variously as Dirichlet or Voronoi tessellations or Theissen polygons.

In Figure 4.14(a), we have a set of points of known height. We can draw lines from A to all neighboring points and construct the perpendicular bisector of each of the lines AB, AC, AD, and AE. Two of these (*wp* and *ps* in Figure 4.14(b)), together with the bounding area, mark out a polygon in which all the points are closer to A than to any of the other points B, C, D, and E. That is because all the perpendicular bisectors separate out areas one side of which is closer to A and the other is closer to the second generating point.

This process can be extended to all the other points creating a series of polygons within which all parts of the area are closer to the generating point than to any other. Thus, for example, in Figure 4.14(c), the lines of the bounding rectangle together with *w* to *p* to *q* to *r* to *v* define a polygon around E, which is the area within which all points lie closer to E than to A, B, C, D, etc.

Such polygons are known as Theissen polygons and are of particular interest in certain types of statistical mapping. The lines connecting nearby points in the final creation of these polygons form a series of triangles known as the Delaunay triangles. They are deemed to be the "best-shaped" triangles, and are the nearest to being equilateral that can be constructed from the original points. These triangles are shown in Figure 4.14(c) by the darker lines.

4.5 THE SPHERE AND THE ELLIPSE

Let us now return to the geometry of the circle since it is important in two areas of geomatics — the study of position fixing and in the representation of the shape of the earth, which is approximately spherical.

A plane intersects a sphere in a circle; if the plane passes through the center of the sphere, as in the case of the meridians of longitude on the Earth, the line on the surface is said to be a *great circle*. Other circles such as parallels of latitude (except the special case of the equator) are called *small circles*.

The geometry of any figure on the surface of a sphere differs from that on a plane because triangles become spherical triangles and their three angles no longer add up to 180°. Thus, in Figure 4.15, angle $NG'P = 90° =$ angle NPG'.

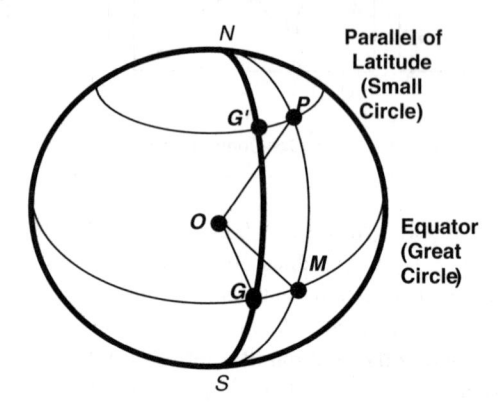

Figure 4.15 Great and small circles.

The angle $G'NP$ at the north pole is the difference in longitude between the meridian or great circle through Greenwich and the meridian through P; this means that the three angles of the triangle $G'NP$ are in excess of 180°. When the sides of the triangles are defined by great circles, the amount by which the sum of the three angles exceeds 180° is called the *spherical excess*. Computations involving spherical triangles will be considered in Chapter 5 on trigonometry.

A sphere is only an approximation to the shape of the earth; in reality, an *ellipsoid* (obtained by rotating an ellipse about its shorter or "minor" axis) gives a better representation. An *ellipse* (Figure 4.16) is a closed figure shaped like a compressed circle that is symmetrical about two axes known as the *major axis* and the *minor axis*. The circle that is formed using the major axis as the diameter is known as the auxiliary circle. The radius of this circle is known as the semimajor axis and is normally denoted 'a' while the length of the semiminor axis is denoted 'b.' In the geometry of the ellipse compared with the auxiliary circle, the distances in the x-direction remain the same but the distances in the y-direction are compressed or scaled down so that P' on the circle becomes P on the ellipse.

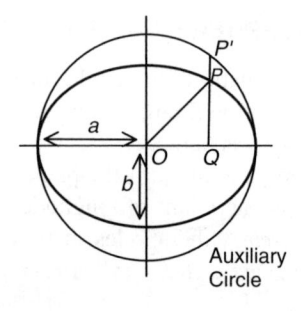

Figure 4.16 Ellipse and auxiliary circle.

4.6 SECTIONS OF A CONE

The ellipse and circle are two of four special second-order curves known as *conic sections*, the other being known as the parabola and the hyperbola. They take the form of

$$ax^2 + by^2 + cxy + dx + ey + f = 0$$

where a, b, c, d, e, and f are constants. For a circle, $a = b$ and $c = 0$. Geometrically, they can be derived from a line known as the *directrix* shown as the line CD in Figure 4.17. F_1 and F_2 are two points on the major axis known as the foci, each *focus* being the same distance either side of the center O on the major axis. Q is the foot of the perpendicular from P onto the line CD so that PQ is parallel to the major axis.

The ellipse has the following characteristics:

1. For all conics, the ratio between the distances F_2P and PQ is a constant that is known as the *eccentricity* 'e' (not to be confused with the Euler number 'e' that equals 2.718 …). For an ellipse, the value of 'e' is such that $0 < e < 1$.
2. If the origin O has coordinates $(0, 0)$ then for the ellipse F_1 is located at the point $(-ae, 0)$ and F_2 at $(+ae, 0)$, where 'a' is the length of the semimajor axis and 'e' is the eccentricity.
3. The directrix is located at a distance $\left(\dfrac{a}{e}\right)$ from the center of the ellipse. It thus cuts the major axis at the point $\left(\dfrac{a}{e}, 0\right)$.
4. The sum of the distances $F_1P + PF_2$ is a constant for the ellipse.
5. $ae = \sqrt{(a^2 - b^2)}$ or $e = \sqrt{(1 - b^2/a^2)}$.
6. The area of an ellipse is πab.
7. The length of the circumference of an ellipse is a complicated thing to calculate.
8. If P is a point on the ellipse with coordinates (x, y) then

$$x^2/a^2 + y^2/b^2 = 1$$

These relationships will be explained more fully in Chapter 8 after we have discussed trigonometric functions. The circle can be regarded as a special case of an ellipse in which $a = b = r$ is the radius of the circle and $x^2/r^2 + y^2/r^2 = 1$. The

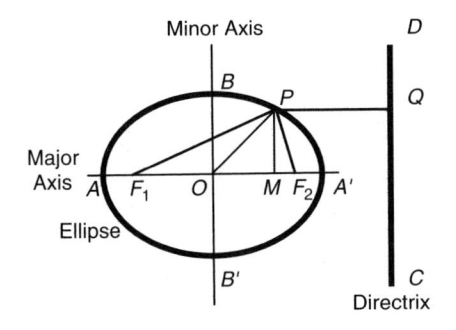

Figure 4.17 An ellipse and its directrix.

eccentricity $e = 0$ and the directrix is located a distance $(a/0)$ from the center, which means that it is an infinite distance away.

For the ellipse $0 < e < 1$. When $e = 1$ we have $FP = PQ$ (see Figure 4.18) and instead of being gently rounded the ellipse becomes infinite in size. The curve is known as a *parabola*. It has only one focus and its directrix is symmetrically placed on the other side of the origin (so that if F is at $(a, 0)$ the directrix passes through $(-a, 0)$). A simple example of a parabola is the function $y^2 = mx$.

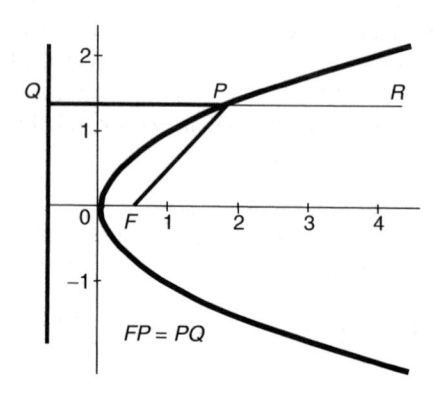

Figure 4.18 A parabola.

The parabola has only one focus and the tangent to the curve at the point P exactly bisects the lines FP and QP. This means that the line PR in Figure 4.18 is parallel to the axis and also at the same angle to the tangent at P as is the line PF. Hence, if a light source were placed at F and the inside of the parabolic surface were perfectly reflecting, then a beam of light from F would emerge parallel to the major axis. Conversely, a parabolic mirror can focus a parallel beam of light or other forms of radiation, for instance, from the sun, onto a single point (F).

If the eccentricity 'e' has a value greater than 1 then the shape of the resulting curve is a *hyperbola*. It has an equation of the form $x^2/a^2 - y^2/b^2 = 1$. It consists of two branches formed by a pair of intersecting lines known as the asymptotes (AB and CD in Figure 4.19) to which it approaches more and more closely but only reaches them at infinity. The slope of these asymptotes is given by the equations

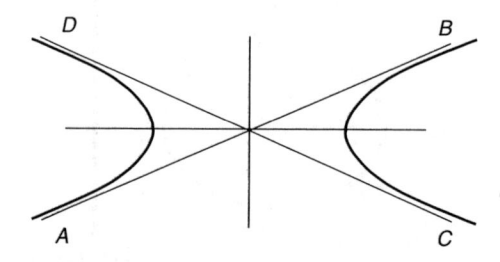

$$y = + (b/a)x \quad \text{and} \quad y = - (b/a)x.$$

Figure 4.19 A hyperbola with its asymptotes.

THE GEOMETRY OF COMMON SHAPES

The circle, ellipse, parabola, and hyperbola all have solid equivalents (the sphere, the ellipsoid, the paraboloid, and the hyperboloid). They are all quadratic forms and in their plane (i.e., 2D) form they are cross-sections of a 3D cone (Figures 4.20 and 4.21).

Consider two intersecting lines. These form a plane. Draw one of the lines of bisection and then in three dimensions rotate the original lines around this bisector to generate a double cone (one cone stuck upside down on top of the other). Cut this solid along a plane perpendicular to the axis of rotation and you have a circle; cut it at an angle less than the slope of the sides of the cone and the cross-section will be an ellipse. If the section of the double cone is cut parallel to the edges of the cone, then if the cone is infinitely large the section will be infinitely large and be a parabola. If, however, the section is greater it will cut the double cone twice (top and bottom) in a hyperbola. We will revisit conic sections in Chapter 9 in the context of map projections.

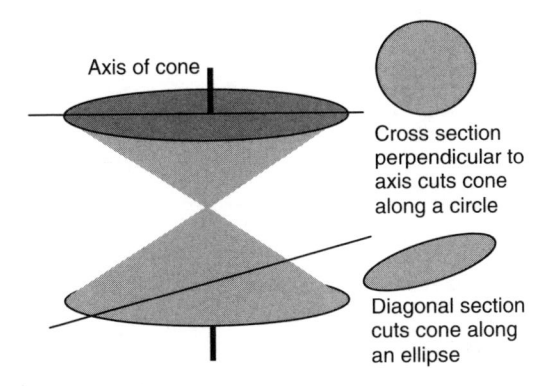

Figure 4.20 Sections of a cone-circle and ellipse.

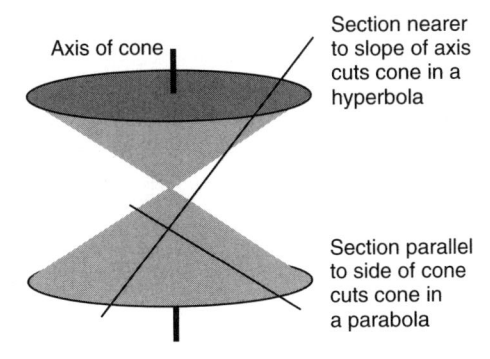

Figure 4.21 Sections of a cone-parabola and hyperbola.

CHAPTER 5

Plane and Spherical Trigonometry

CONTENTS

5.1   Basic Trigonometric Functions 59
5.2   Obtuse Angles 63
5.3   Combined Angles 65
5.4   Bearings and Distances 67
5.5   Angles on a Sphere 72

5.1 BASIC TRIGONOMETRIC FUNCTIONS

Trigonometry is concerned with ratios between the sides and angles of triangles. Although at its simplest it is concerned with right-angled triangles on a plane surface, its applications extend to many areas of geomatics including calculations on curved surfaces such as that of the Earth. Trigonometry is used extensively in surveying and navigation.

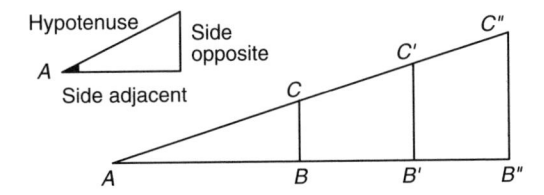

Figure 5.1 Similar right-angled triangles.

In Chapter 3, Figure 3.6 showed similar triangles that have a common shape but differ in scale. In Figure 5.1, the triangles ABC, $AB'C'$, and $AB''C''$ are all the same shape (they have the same angles) but they differ in size or scale. In these triangles, the ratio

$$\frac{BC}{BA} = \frac{B'C'}{B'A} = \frac{B'A}{C''A} = \text{``Side Opposite over Side Adjacent''}$$

Given the fixed angle *A*, in any right-angled triangle, the ratio *BC/BA* is constant. This is called the *tangent* of angle *A* or tan *A*. Similarly, the ratio *BC/AC* is constant = "Side Opposite over Hypotenuse." This is called the *sine* of angle *A* or sin *A*. Likewise, the ratio *AB/AC* is a constant = "Side adjacent over Hypotenuse." It is called the *cosine* of angle *A* or cos *A* (Box 5.1).

Box 5.1 — Functions of Angle A

sin *A*, cos *A*, and tan *A* are functions of the angle *A*. Their values can be calculated using special formulae (Box 6.4) or be obtained from tables, pocket calculators, etc.

As an example for $A = 35°$ from tables with 7 decimal places

$$\sin 35° = 0.5735764$$

$$\cos 35° = 0.8191520$$

$$\tan 35° = 0.7002075$$

Note

$$(\sin 35)*(\sin 35) = (\sin 35)^2 = 0.3289899$$

$$(\cos 35)*(\cos 35) = (\cos 35)^2 = 0.6710101$$

Thus, $\sin^2(35) + \cos^2(35) = 1$. Likewise,

$$\sin 35/\cos 35 = 0.5735764/0.8191520$$

$$= 0.7002075$$

$$= \tan 35$$

The ratios sin, cos, and tan are not independent. Given the theorem by Pythagoras

$$BC^2 + AB^2 = AC^2$$

Dividing both sides by AC^2, we obtain $(BC/AC)^2 + (AB/AC)^2 = 1$, or

$$(\sin A)^2 + (\cos A)^2 = 1$$

Also

$$(\sin A/\cos A) = \left\{\frac{BC}{AC}\right\} \bigg/ \left\{\frac{AB}{AC}\right\} = \frac{BC}{AB} = \tan A$$

Hence

$$\tan = \frac{\sin}{\cos}$$

Box 5.2 — Basic Relationships

Sine = sin = side opposite/hypotenuse
Cosine = cos = side adjacent/hypotenuse
Tangent = tan = side opposite/side adjacent

$$\sin/\cos = \tan$$

$$(\sin A)^2 + (\cos A)^2 = 1$$

$$\text{cosec} = 1/\sin;\ \sec = 1/\cos \text{ and } \cot = 1/\tan$$

It is helpful sometimes to deal with the reciprocal ratios AC/AB and AC/BC. These (and the reciprocal of tan A) have special names. 1 divided by sine is called *cosecant* or *cosec*; 1 divided by cosine is the *secant* or *sec*; and 1 divided by tangent = *cotangent* or *cot* (Box 5.2). By convention, just as $x*x = x^2$, so $\sin A * \sin A = (\sin A)^2$ and is written as $\sin^2 A$. Likewise, $\cos A * \cos A = \cos^2 A$, etc. However, although $1/x = x^{-1}$, $1/\sin A$ is NOT written as $\sin^{-1} A$ since this has a reserved meaning, which is "the angle whose value is sine A." This in turn is called *arc-sine A* or arc-sin A. Cos^{-1} means "the angle whose value is cosine A." It is called *arc-cos*. Similarly, $\tan^{-1} = $ *arc-tan*, $\text{cosec}^{-1} = $ *arc-cosec*, $\sec^{-1} = $ *arc-sec*, and $\cot^{-1} = $ *arc-cot*.

In Figure 5.1, ABC is a right-angled triangle and the angle at C (the angle ACB) has as its cosine the ratio BC/AC. Thus, cos C is the same as sin A. This only happens when the angle $C = 90° - A$. In general, however, $\sin A = \cos(90° - A)$; $\cos A = \sin(90° - A)$; hence, $\tan A = \cot(90° - A)$.

We can always convert non-right-angled triangles into ones with right angles as shown in Figure 5.2. Even though the triangle ABC has no right angle there is still a value for "sin A" since this is dependent on the size of the angle A not the shape or size of the triangle. We can create a right-angled triangle (Figure 5.2) by drawing the perpendicular line from C onto the horizontal line AB so that angle $ANC = 90°$. We represent the sides of the triangle with lowercase letters (with $BC = a$, $CA = b$, and $AB = c$ and the height $CN = h$, known as the altitude) and the angles with uppercase letters A, B, C.

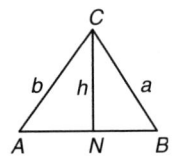

Figure 5.2 An altitude of a triangle.

In Figure 5.2, $\sin A = CN/AC = h/b$ or $h = b \sin A$. Also $\sin B = CN/BC = h/a$ or $h = a \sin B$. Thus,

$$b \sin A = a \sin B \text{ or } \frac{\sin A}{a} = \frac{\sin B}{b}$$

By constructing the perpendicular from B, we can also show that $(\sin A)/a = (\sin C)/c$. Thus for any triangle ABC with sides a, b, c

$$\frac{\sin A}{a} = \frac{\sin B}{b} = \frac{\sin C}{c}$$

This is known as the *sine formula* for triangle ABC. Note also that the

area of the triangle $= (1/2)$ base$*$height $= (1/2)$ ch

$$= (1/2) \, bc \sin A = (1/2) \, ca \sin B = (1/2) \, ab \sin C$$

Further, in Figure 5.2 $\cos A = AN/AC$. Thus, $AN = AC \cos A = b \cos A$. Using Pythagoras, $AN^2 + CN^2 = AC^2$, or $CN^2 = AC^2 - AN^2 = b^2 - b^2 \cos^2 A$. Likewise in triangle BNC where $NB = (c - AN) = (c - b \cos A)$. We have

$$CN^2 = BC^2 - BN^2 = a^2 - (c - b \cos A)^2$$
$$= a^2 - \{c^2 - 2bc \cos A + (b \cos A)^2\}$$
$$= a^2 - c^2 + 2bc \cos A - b^2 \cos^2 A$$
$$= b^2 - b^2 \cos^2 A \text{ (from the above relations)}$$

Hence $a^2 - c^2 + 2bc \cos A = b^2$ or $2bc \cos A = b^2 + c^2 - a^2$. Thus,

$$\cos A = (b^2 + c^2 - a^2)/2bc$$

Similarly,

$$\cos B = (c^2 + a^2 - b^2)/2ca, \cos C = (a^2 + b^2 - c^2)/2ab$$

These relationships are known as the *cosine formule* (Box 5.3).

Box 5.3 — Sine and Cosine Formulae for any Plane Triangle

1. $(\sin A)/a = (\sin B)/b = (\sin C)/c$

2. $\cos A = (b^2 + c^2 - a^2)/2bc$

   $\cos B = (c^2 + a^2 - b^2)/2ca$

   $\cos C = (a^2 + b^2 - c^2)/2ab$

Note

area of the triangle $= (1/2) \, bc \sin A = (1/2) \, ca \sin B = (1/2) \, ab \sin C$

5.2 OBTUSE ANGLES

So far, we have assumed that the angles A, B, and C are less than or equal to a right angle. If we think of the case where ABC has B as a right angle, then the larger the A becomes, the smaller the angle C (Figure 5.3).

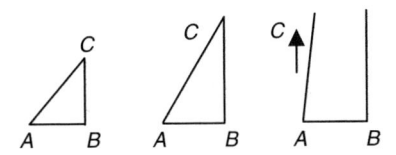

Figure 5.3 Towards a right-angle.

As A approaches 90°, angle C approaches zero while BC and AC become parallel and of similar length. In the limit, sin 90° = 1 while cos 90° = 0. Since tan 90° = sin 90°/cos 90° = 1/0 and we cannot divide by zero it means that as A approaches 90°, tan A becomes considerably larger until it becomes infinitely large. But what happens for angles over 90°? The sine formula and the cosine formula apply to triangles where there is a fixed relationship between the angles A, B, and C. The functions sin, cos, and tan give values that relate to the size of the angle not the shape or size of a triangle. We can look at them in a different way. Consider a circle of unit radius (Figure 5.4) and center A and a point P in the quadrant or quarter of the circle between the +x- and +y-axis:

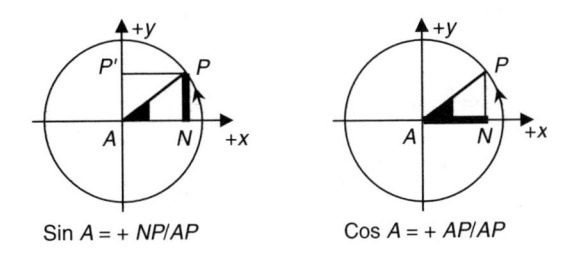

Sin A = + NP/AP   Cos A = + AP/AP

Figure 5.4 A circle with unit radius.

If sin A = NP/AP and AP = 1 (unit radius), then the length of NP = sin A. AP' is the projection of P onto the +y-axis and also = sin A. P' is on the +y-side of the origin. Likewise, the length AN = cos A. This is the projection of P onto the +x-axis. As P moves anticlockwise around the circle the angle that AP makes with the +x-axis increases until at 90° the projection of AP onto the +y-axis reaches the point where NP = AP' = 1. As the angle increases further, the length of NP becomes shorter. It still has the value of sin A. On the other hand, AN has moved from 1 down to 0 and is now on the negative side of the axis. Comparing Figures 5.4 and 5.5, with P in Figure 5.5 the mirror image of what it was in Figure 5.4, we find that the lengths AP' (the projection of P onto the +y-axis) are exactly the same; but AN, the projection of P on the +x-axis, has become negative in the second case.

For angles where $90° \leq A \leq 180°$

$$\sin A = + \sin(180 - A)$$

But

$$\cos A = - \cos(180 - A)$$

Hence, since $\tan = \sin/\cos$

$$\tan A = - \tan(180 - A)$$

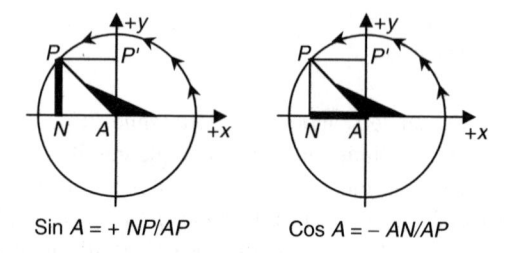

Sin $A = +$ NP/AP   Cos $A = -$ AN/AP

Figure 5.5 Angles in the second quadrant.

A similar argument applies to angles greater than 180° and less than 270° in what is called the third quadrant. In Figure 5.6(a), both AN and NP are negative

$$\sin A = -\sin(A - 180), \quad \cos A = -\cos(A - 180), \quad \tan A = +\tan(A - 180)$$

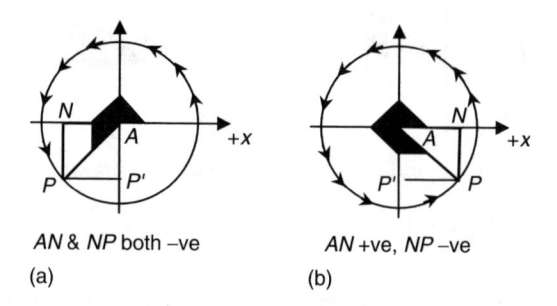

AN & NP both −ve   AN +ve, NP −ve
(a)   (b)

Figure 5.6 Angles in the (a) third and (b) fourth quadrants.

And for the fourth quadrant (Figure 5.6(b)), where AN has again become positive but NP is still negative,

$$\sin A = -\sin(360-A), \quad \cos A = +\cos(360-A), \quad \tan A = -\tan(360-A)$$

In particular, $(360-A)$ is a complete turn less A $(=-A)$. Thus,

$$\sin(-A) = -\sin A, \quad \cos(-A) = +\cos A, \quad \tan(-A) = -\tan A$$

PLANE AND SPHERICAL TRIGONOMETRY

Also, since $\sin A = \cos (90 - A)$

$$\sin (90 + A) = \cos \{90 - (90 + A)\} = \cos (-A) = + \cos A$$

And $\cos (A) = \sin (90 - A)$, then

$$\cos (90 + A) = \sin \{90 - (90 + A)\} = \sin (-A) = -\sin A$$

So

$$\tan (90 + A) = -\cot A$$

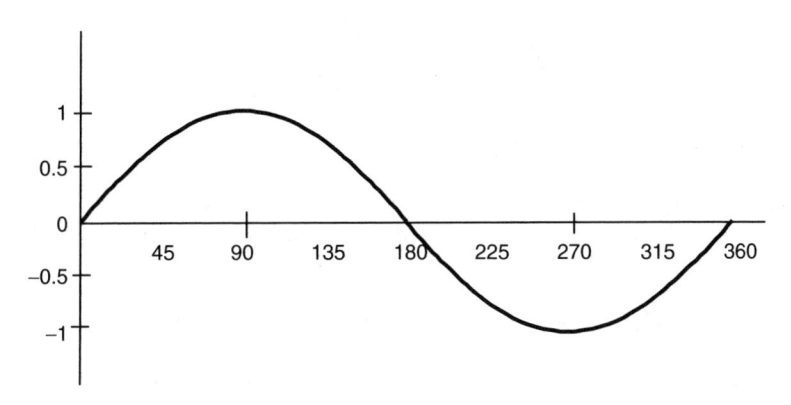

Figure 5.7 The cycle of values of sin A.

Figure 5.7 shows the values of sin A for angles between 0 and 360°. The cycle is repeated every 360°. The curve representing the function cos A has exactly the same shape but a quarter of a cycle out of phase. If 90° is subtracted from the numbers on the horizontal scale on the graph in Figure 5.7, the resulting curve would be for cos A. The summary is given in Table 5.1.

Table 5.1 Signs of Sines and Cosines

Range	Sine	Cosine	Tangent
Values of Sine, Cosine, and Tangent			
0–90	+	+	+
90–180	+	−	−
180–270	−	−	+
270–360	−	+	−

5.3 COMBINED ANGLES

In later chapters, we will make use of this information; but for the present, there are two more relationships that need to be established. Consider two adjacent angles A and B in Figure 5.8 defined by the lines OS, OR, and OQ. From the point S (anywhere along the line OS — scale does not matter), draw the perpendicular

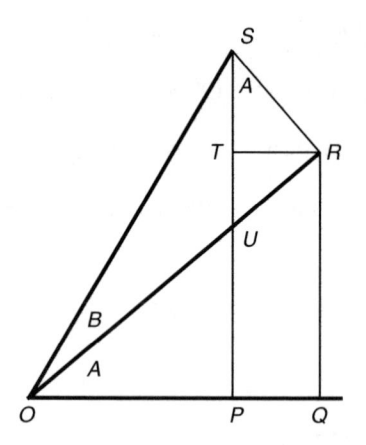

Figure 5.8 Combining adjacent angles.

from S onto OQ at P cutting OR at U. Also, choose R as the point that gives angle $ORS = 90°$ and draw the other horizontal or perpendicular lines RT and RQ. None of this alters the angles A and B.

$$\angle OUP = 90 - A; \angle OUS = 180 - (90 - A) = 90 + A$$

Hence, $\angle SUR = 90 - A$. Also $\angle SUR = \angle OUP$ (Rule 4.2 in Chapter 4). Since $\angle SRU = 90$ by construction, $\angle USR = A$.

$$\sin B = \frac{RS}{OS}, \quad \cos B = \frac{OR}{OS}, \quad \sin A = \frac{QR}{OR}, \quad \cos A = \frac{OQ}{OR}$$

$$\sin(A+B) = \frac{PS}{OS} = \frac{PT + TS}{OS} = \frac{QR + TS}{OS}$$

since $PT = QR$

$$\sin(A+B) = \frac{QR}{OS} + \frac{TS}{OS} = \frac{QR}{OR} * \frac{OR}{OS} + \frac{TS}{SR} * \frac{SR}{OS}$$

by dividing the top and bottom of the first term by OR and the second term by SR. Hence,

$$\sin(A+B) = \sin A \cos B + \cos A \sin B$$

Similarly,

$$\cos(A+B) = \frac{OP}{OS} = \frac{OQ - PQ}{OS} = \frac{OQ - RT}{OS}$$

$$= \frac{OQ}{OS} - \frac{RT}{OS} = \frac{OQ}{OR} * \frac{OR}{OS} - \frac{RT}{SR} * \frac{SR}{OS}.$$

Hence,

$$\cos(A + B) = \cos A \cos B - \sin A \sin B$$

The results are summarized in Box 5.4.

PLANE AND SPHERICAL TRIGONOMETRY 67

Box 5.4 — Combined Angles

$$\sin(A + B) = \sin A \cos B + \cos A \sin B$$
$$\cos(A + B) = \cos A \cos B - \sin A \sin B$$

Since $\sin(-B) = -\sin B$ and $\cos(-B) = +\cos B$

$$\sin(A - B) = \sin A \cos B - \cos A \sin B$$
$$\cos(A - B) = \cos A \cos B + \sin A \sin B$$
$$\sin 2A = \sin(A + A) = 2 \sin A \cos A$$
$$\cos 2A = \cos(A + A) = \cos^2 A - \sin^2 A$$

And since $\cos^2 A + \sin^2 A = 1$

$$\cos 2A = 1 - 2 \sin^2 A = 2\cos^2 A - 1$$

Alternatively,

$$\sin A = 2 \sin(A/2) \cos(A/2)$$
$$\cos A = \cos^2(A/2) - \sin^2(A/2)$$
$$\text{or } \cos A = 1 - 2 \sin^2(A/2) = 2\cos^2(A/2) - 1$$

5.4 BEARINGS AND DISTANCES

The formulae given in Box 5.4 are necessary and sufficient for the solution of a wide range of problems in surveying and mapping. Before considering some examples, we need to clarify the difference between a mathematician's approach and that of a surveyor. Traditionally, in mathematics, angles are measured from the horizontal anticlockwise. Surveyors and navigators start with the north and measure angles and bearings clockwise (Figure 5.9).

Mathematics | Surveying

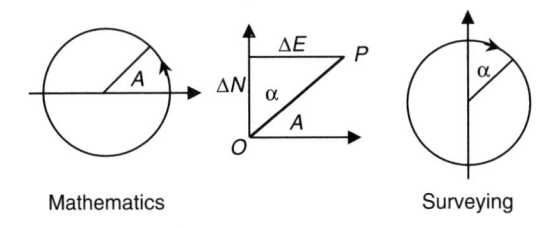

Figure 5.9 Angle and bearing measurements.

When dealing with bearings, the Greek alphabet is often used. Bearings are angles and conform to all the rules that affect the angles A, B, and C that we have been discussing. Thus, in the center triangle of Figure 5.9, the value $\tan \alpha = \Delta E/\Delta N$ while $\tan A = \Delta N/\Delta E$ (as shown in Box 5.5).

> **Box 5.5 — Bearings and Distances from Coordinates**
>
> If point O in Figure 5.10 is $(2624.81E, 3427.64N)$
> and P $(3056.61E, 4058.18N)$, then $\Delta E = 431.80$, $\Delta N = 630.54$
>
> $$\Delta E/\Delta N = 0.6848098, \tan^{-1}(0.6848098) = 34.40373$$
>
> $$\alpha = 34.40373° = 34° \ 24 \min 13.4 \sec = \text{bearing from } O \text{ to } P$$
>
> Sin $\alpha = \Delta E/OP$. Hence
>
> $$OP = \Delta E \operatorname{cosec} \alpha = 431.8 * 1.76985 = 764.22$$
>
> Cos $\alpha = \Delta N/OP$. Hence
>
> $$OP = \Delta N \sec \alpha = 630.54 * 1.21201 = 764.22$$
>
> This verifies the calculation.

The accuracy of the process of calculating bearings and distances from coordinates depends upon the number of significant figures that are used throughout the computation, as illustrated in Box 5.6.

> **Box 5.6 — Bearings and Distances into Coordinates**
>
> If O in Figure 5.10 has coordinates $(2624.81E, 3427.64N)$
> and the distance OP = 764.22 and bearing $\alpha = 34° \ 24'$
>
> $$\Delta E = OP \sin \alpha = 764.22 * 0.5649670 = 431.76$$
>
> $$\Delta N = OP \cos \alpha = 764.22 * 0.8251135 = 630.57$$
>
> These figures differ slightly from the values given in Box 5.5. This is because the bearing should in fact be
>
> $$\alpha = 34° \ 24' 13.4''$$
>
> giving
>
> $$\Delta E = 764.22 * 0.5650206 = 431.80$$
>
> $$\Delta N = 764.22 * 0.8250766 = 630.54$$

Taking the relationships a stage further, consider Figure 5.10 where the point Q forms a triangle with OP. Q is to the right of the line from O to P, and the measured angles are α and β. Let the coordinates of O be (E_O, N_O) P (E_P, N_P) and

PLANE AND SPHERICAL TRIGONOMETRY

Q (E_Q, N_Q). If the bearing from O to P is ϕ then the bearing $OQ = \phi + \alpha$. The bearing P to $O = 180 + \phi$; thus, the bearing $PQ = 180 + \phi - \beta$. Angle $Q = 180 - (\alpha + \beta)$ and

$$\sin Q = \sin\{180 - (\alpha + \beta)\} = \sin(\alpha + \beta) = \sin\alpha \cos\beta + \cos\alpha \sin\beta$$

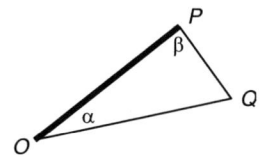

Figure 5.10 Fixing points from observed angles.

Using the sine formula, $OQ = OP \sin\beta/\sin Q$. As we have seen, $OP = (E_P - E_O)/\sin\phi = (N_P - N_O)/\cos\phi$.

$$E_Q = E_O + OQ \sin(\phi + \alpha) = E_O + OQ \sin\phi \cos\alpha + OQ \cos\phi \sin\alpha$$

$$= E_O + \{OP \sin\beta \sin\phi \cos\alpha + OP \sin\beta \cos\phi \sin\alpha\}/[\sin\alpha \cos\beta + \cos\alpha \sin\beta]$$

By substituting for OP, we obtain the relations shown in Box 5.7 and 5.8. If Q were on the other side of the line OP, then the two last terms in the numerator of both expressions in Box 5.7 would be of opposite sign. Thus, references to P and O would be reversed.

Box 5.7 — Coordinates from Observed Angles

If point Q is to the right of the line from O to P as in Figure 5.10, then

$$E_Q = \frac{E_P \cot\alpha + E_O \cot\beta + N_P - N_O}{\cot\alpha + \cot\beta}$$

$$N_Q = \frac{N_P \cot\alpha + N_O \cot\beta - E_P + E_O}{\cot\alpha + \cot\beta}$$

As an example of the use of trigonometric functions in geomatics, consider the case of traversing, a technique commonly used in land surveying. It entails measuring a series of angles and distances.

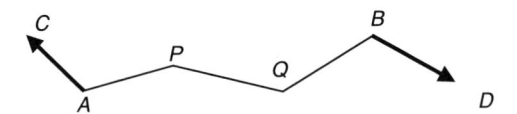

Figure 5.11 A traverse.

> **Box 5.8 — Computing a Point from Two Observed Angles**
>
> Using the previous data from Box 5.5 and in Figure 5.10:
> Let O have coordinates (2624.81E, 3427.64N) and let P have coordinates (3056.61E, 4058.18N). Then
>
> $$\Delta E = 431.80, \quad \Delta N = 630.54$$
>
> Let $\alpha = 20°$ and $\beta = 60°$. Then, $\cot \alpha = 2.7474774$, $\cot \beta = 0.5773503$, $\cot \alpha + \cot \beta = 3.3248277$.
>
> $$E_Q = \frac{E_P \cot \alpha + E_O \cot \beta + N_P - N_O}{\cot \alpha + \cot \beta}$$
>
> $$= (8397.967 + 1515.435 + 630.54)/3.3248277$$
>
> $$= 3171.27$$
>
> $$N_Q = \frac{N_P \cot \alpha + N_O \cot \beta - E_P + E_O}{\cot \alpha + \cot \beta}$$
>
> $$= (11149.758 + 1978.949 - 431.80)/3.3248277$$
>
> $$= 3818.82$$
>
> Hence, the coordinates of Q are (3171.27, 3818.82)

The traverse would start at a known point (A in Figure 5.11) and end at another known point (B) in order to confirm the accuracy of the work. The bearings and distances from A to P to Q to B would be derived from observations so that the coordinates of the two new points P and Q could be calculated. The measurement of the distances AP, PQ, and QB would be relatively straightforward, especially with modern electronic distance measuring devices.

The measurement of bearings is more problematic since the bearing of AP is an angle relative to North; but in what direction is North? The solution is to have another known point (here C), calculate the bearing of AC, measure the angle between AC and AP clockwise, and hence derive the bearing of AP. The bearing PQ is then derived by measuring the angle at P.

Because all measurements and calculations are prone to error, either through human mistakes or the accumulation of small inaccuracies, it is necessary to verify the positions by some independent means. This is done by measuring the angle at Q, the distance QB, and a check angle at B to another known point (D).

In the calculation shown in Table 5.2, the given values are in bold; there are three observed distances and four observed angles from which the coordinates of the two new points have been derived. Note that values that do not require adjustment have

Table 5.2 Traverse Calculation

Station	Distance	Angle	Bearing	Sin Bearing	Cos Bearing	Δ East	Δ North	East	North
C								5663.28	13794.22
	497.68		141 21 02	0.6245538	−0.7809818	310.83	−388.68		
A		108 17 08						5974.11	13405.54
	498.93		69 38 10	0.9375015	0.3479813	467.75	173.62		
P		228 33 27						6441.86	13579.16
	318.95		118 11 37	0.8813561	−0.4724525	281.11	−150.69		
Q		108 44 11						6722.97	13428.47
	399.55		46 55 48	0.7305199	0.6828914	291.88	272.85		
B		246 54 02						7014.85	13701.32
	750.38		113 49 50	0.9147443	−0.4040332	686.41	−303.18		
D								7701.26	13398.14

been used; normally in a traverse there will be a need to apply corrections to ensure exact mathematical consistency. The theory of errors is explored from a statistical perspective in Chapters 10 and 11.

5.5 ANGLES ON A SPHERE

In practice, the world is not flat and we must extend some of the ideas above to what happens on a curved surface. For the present, we shall assume that for practical purposes the world is a sphere. As explained in Chapter 4, on a sphere any plane that passes through the center of the sphere cuts the surface along a line that is called a great circle. Thus, assuming that the Earth is a sphere, the Equator and all meridians of longitude are great circles. The shortest distance between any two points on the surface of a sphere follows the line of a great circle.

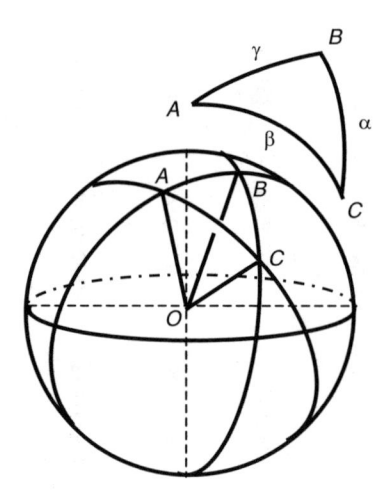

Figure 5.12 The spherical triangle.

A triangle on the surface of a sphere that has each of its arcs as part of a great circle is called a *spherical triangle*. Thus, if O is the center of the sphere in Figure 5.12, then the plane OAB cuts the surface along the arc AB, which is part of a great circle; this is similar for BC and CA. The length of the arc AB is the radius of the sphere multiplied by the angle AOB in radians. This is usually expressed in letters from the Greek alphabet — for example, π radians for 180°. The angle on the surface (for instance BAC) is written in normal text — here as A. For a spherical triangle ABC the angles would then be A, B, and C, while the side opposite A (i.e., BC) would be α (alpha), CA would be β (beta), and AB would be γ (gamma).

Consider a sphere of radius R (Figure 5.13). Let P be the foot of the perpendicular from A onto the plane OBC; let the line PS be perpendicular to OB and let PQ be perpendicular to OC. Triangles APS and APQ both have right angles at P since AP is perpendicular to the plane that contains all the points $OQCBS$. Furthermore, since the points A, S, and P form a plane that is at right angles to the line OB, it

PLANE AND SPHERICAL TRIGONOMETRY

follows that the line AS (not drawn in Figure 5.13) must be at right angles to the line OB or that ASO is a right-angled triangle with the angle at $S = 90°$.

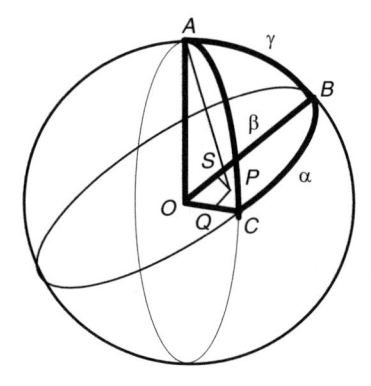

Figure 5.13 Spherical angles.

It also follows that the plane ASP is parallel to the plane that touches the sphere at B, known as the tangent plane at B since both are at right angles to the line OB, which is a radius of the sphere. The angle ASP must therefore be the same as the angle ABC at the surface. Thus, $\angle ASP = B$. Finally, the $\angle AOS = \angle AOB = \gamma$. Similarly, $\angle APQ = 90° = \angle AQO$. Also, $\angle AQP = C$ and $\angle AOQ = \beta$. The relationships are shown in Figure 5.14.

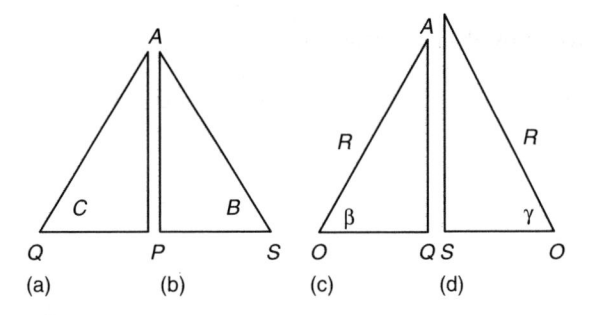

Figure 5.14 The sine formula for spherical triangles.

All this may seem difficult to visualize until we separate out the triangles. Then, we have from Figure 5.14(c), $AQ = OA \sin \beta = R \sin \beta$ and from Figure 5.14(d), $AS = OA \sin \gamma = R \sin \gamma$. Combining these with Figures 5.14(a) and (b), we have from Figure 5.14(a), $AP = AQ \sin C = R \sin \beta \sin C$ and from Figure 5.14(b), $AP = AS \sin B = R \sin \gamma \sin B$. Hence, $\sin \beta \sin C = \sin \gamma \sin B$ or

$$\sin B / \sin \beta = \sin C / \sin \gamma.$$

And, similarly, each of these ratios can be shown to be equal to $\sin A / \sin \alpha$. Thus,

$$\frac{\sin A}{\sin \alpha} = \frac{\sin B}{\sin \beta} = \frac{\sin C}{\sin \gamma}$$

This is known as the *sine formula* for a spherical triangle.

Taking this a step further, in the plane OBC as shown in Figure 5.13, draw SU perpendicular to OC and PT perpendicular to SU. This is shown in Figure 5.15 where $\angle BOC = \alpha$ and hence $\angle OSU = 90° - \alpha$. Since PS is perpendicular to OB, $\angle PST = \alpha$

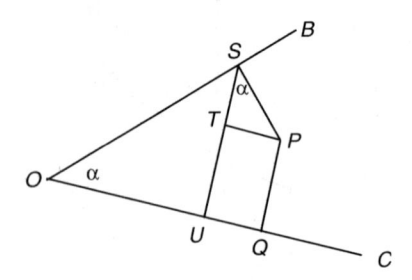

Figure 5.15 The cosine formula for spherical triangles.

From Figure 5.14, $AS = R \sin \gamma$ and $SP = R \sin \gamma \cos B$; also, $OQ = R \cos \beta$ and $OS = R \cos \gamma$. From Figure 5.15, $TP = SP \sin \alpha = R \sin \alpha \sin \gamma \cos B$. Also $TP = UQ$ (since $PTUQ$ is a rectangle) $= OQ - OU = R \cos \beta - OS \cos \alpha = R \cos \beta - R \cos \gamma \cos \alpha$.
Hence,

$$R \sin \alpha \sin \gamma \cos B = R \cos \beta - R \cos \gamma \cos \alpha$$

On dividing through by R and rearranging

$$\cos \beta = \cos \gamma \cos \alpha + \sin \gamma \sin \alpha \cos B$$

Similarly,

$$\cos \gamma = \cos \alpha \cos \beta + \sin \alpha \sin \beta \cos C,$$

$$\cos \alpha = \cos \beta \cos \gamma + \sin \beta \sin \gamma \cos A$$

These are known as the *cosine formulae* for a spherical triangle (Box 5.9).

Box 5.9 — Sine and Cosine Formulae for Spherical Triangles

1. $\dfrac{\sin A}{\sin \alpha} = \dfrac{\sin B}{\sin \beta} = \dfrac{\sin C}{\sin \gamma}$

2. $\cos \alpha = \cos \beta \cos \gamma + \sin \beta \sin \gamma \cos A$

   $\cos \beta = \cos \gamma \cos \alpha + \sin \gamma \sin \alpha \cos B$

   $\cos \gamma = \cos \alpha \cos \beta + \sin \alpha \sin \beta \cos C$

Box 5.10 — Bearing and Distance from Latitude and Longitude

In Figure 5.16, N is the North Pole. A (40° N, 10° E) and B (55° N, 15° E) differ in longitude by 5°. Angle $AN = 50°$, $BN = 35°$ and $ANB = 5°$.
From the cosine formula

$$\cos (AB) = \cos 50 \cos 35 + \sin 50 \sin 35 \cos 5$$

$$= 0.5265408 + 0.4377130 = 0.9642538$$

Hence, the angle $AB = 15° 22'$ or 0.2681836 rad. This is the angle AOB subtended by AB at the center. Assuming the radius of the Earth to be $6.4*10^6$ m
The length of $AB = 0.2681836^r * 6.4 * 10^6 = 1716.4$ km.
Using the sine formula,

$$(\sin ANB)/(\sin AB) = (\sin NAB)/(\sin NB)$$

or

$$\sin NAB = \sin NB * \sin ANB / \sin AB = 0.1886573$$

Hence,

$$\text{angle } NAB = 10° 52' = \text{Bearing from } A \text{ to } B$$

Similarly,

$$\sin NBA = \sin NA * \sin ANB / \sin NB = 0.2519627$$

Hence,

$$\text{angle } NBA = 165° 24'$$

$$\text{Bearing } BA = 194° 36'$$

Note that in general, unlike on a plane surface, the direction from A to B on a sphere does not normally differ by 180° from the direction from B to A.

As an example of the application for the sine and cosine formulae, consider two points A and B with the following latitude and longitude (see Figure 5.16):

$$A \text{ (40° N, 10° E) and B (55° N, 15° E)}$$

The angle ANB at the north pole = difference in longitude = 5°. The latitude of A is the angle that OA makes with the equator, which means that the angle NOA = length of arc $AN = 90°$ − latitude of A (known as the colatitude of A) = 50°. Similarly, the length of arc $NB = 35°$ (the colatitude of B). We therefore have a spherical triangle in which there are two sides (*AN* and *BN*) and an included angle (*ANB*). Hence, we can calculate the bearing and distance of *A* to *B*, using the spherical sine and cosine formulae as shown in Box 5.10.

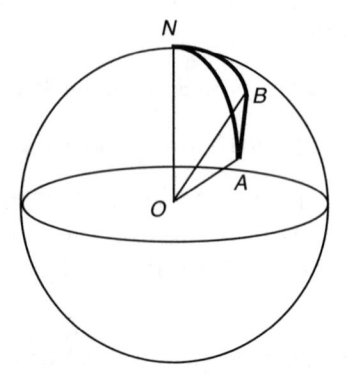

Figure 5.16 Colatitudes.

CHAPTER 6

Differential and Integral Calculus

CONTENTS

6.1 The Basis of Differentiation 77
6.2 Differentiating Trigonometric Functions...81
6.3 Polynomial Functions 83
6.4 Basic Integration 85
6.5 Areas and Volumes 87

6.1 THE BASIS OF DIFFERENTIATION

Calculus is a branch of mathematics, whose origin was the basis of an acrimonious dispute between two great mathematicians — Sir Isaac Newton and Gottfried Leibniz — both of whom claimed to be the discoverer. *Differential calculus* is concerned with the rate at which a function changes while *integral calculus* extends the idea that the sum of a finite number of separate values can be used to form a continuous value as, for instance, in determining the area under a curve.

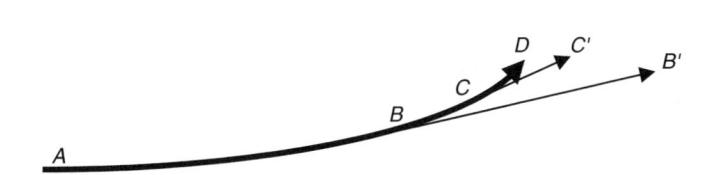

Figure 6.1 Tangents to a curve.

Consider first a function such as the parabola *ABCD*, etc., (Figure 6.1) in which $y = ax^2$ and 'a' is some constant. If we travel from a point *B* with coordinates (x, y) on the curve to a point *C*, the direction in which we are moving along the curve changes from *BB'* to *CC'*. These directions are called the *tangents* at *B* and *C*. They are related to the idea of the tangent of an angle as discussed in Chapter 5 since the slope of the straight line from B to C is the ratio $\delta y/\delta x$ in Figure 6.2 (δ being the Greek letter known as delta and shown here in lower case) and is the tangent of the angle θ (the Greek letter theta) of the slope of the straight line BC.

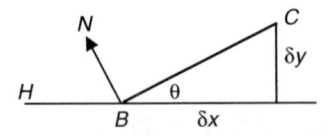

Figure 6.2 The slope and the normal.

Now if the distance marked as 'δx' is small (δ standing for "a small amount of") and if, on moving from B to C along the curve, the y value increases by an equivalent small amount δy then the coordinates of C must be $(x + \delta x, y + \delta y)$. These must also satisfy the basic equation $y = ax^2$.
Hence

$$y + \delta y = a(x + \delta x)^2 = ax^2 + 2ax\delta x + a\delta x^2$$

Since for the point B, $y = ax^2$, it follows that

$$\delta y = 2ax\,\delta x + a\delta x^2 = (2ax + a\delta x)\,\delta x$$

Dividing both sides by δx,

$$\delta y/\delta x = 2ax + a\delta x$$

As δx becomes smaller and smaller so does δy and the ratio between the two, namely $\delta y/\delta x$ tends to the value $= 2ax$. In the limit, we use the ordinary alphabet and say that $dy/dx = 2ax$. This process is known as *differentiation*. Thus, if we differentiate y with respect to x where $y = ax^2$ we obtain $dy/dx = 2ax$. Note that δx means "a little bit of x" but dy/dx means "in the limit the value of $\delta y/\delta x$ when both δy and δx have become very, very small." If we have a function of the form $y = ax^2 + bx + c$, then

$$y + \delta y = a(x + \delta x)^2 + b(x + \delta x) + c$$
$$= ax^2 + 2ax\delta x + a(\delta x)^2 + bx + b\delta x + c$$

or

$$\delta y = 2ax\delta x + a(\delta x)^2 + b\delta x$$

or dividing both sides by δx, $\delta y/\delta x = 2ax + b + a\delta x$.
In the limit as δx tends to zero $dy/dx = 2ax + b$.
We can extend the argument to a cubic and find that if

$$y = ax^3 + bx^2 + cx + e$$

DIFFERENTIAL AND INTEGRAL CALCULUS

Then

$$dy/dx = 3ax^2 + 2bx + c$$

And in general if

$$y = ax^n \quad \text{then} \quad dy/dx = nax^{(n-1)}$$

If we have two functions u and v that are both functions of x and if $y = uv$, then

$$y + \delta y = (u + \delta u)(v + \delta v) = uv + u\delta v + v\delta u + \delta u \cdot \delta v$$

$$\delta y = u\delta v + v\delta u + \delta u \cdot \delta v$$

$$\delta y/\delta x = u\delta v/\delta x + v\delta u/\delta x + \delta u \cdot \delta v/\delta x$$

The expression $\delta u \cdot \delta v$ becomes smaller twice as fast as either δu or δv separately; hence $+ \delta u \cdot \delta v/\delta x$ tends to zero. The result is that

$$dy/dx = u \, dv/dx + v \, du/dx$$

(Note, that if $y = x^2 = x*x$, then $u = v = x$ and $dy/dx = x(dx/dx) + x(dx/dx) = 2x$)

"dy/dx" is the rate of change of the function y with respect to x or, viewed in another way as in Figure 6.1, it is the slope of the curve at the point (x, y). The line that is perpendicular to the tangent (BN in Figure 6.2) is known as the *normal* to the curve. Note that the slope of BN $= \theta + 90°$ and that

$$\tan(\theta + 90°) = \sin(\theta + 90°)/\cos(\theta + 90°) = -\cos\theta/\sin\theta = -\cot\theta = -(\delta x/\delta y)$$

Thus, the tangent of the slope of the normal $= -1$ divided by the tangent of the slope of the curve.

In general, if y is a function of x, we can write this as $y = f(x)$. When we differentiate the function $f(x)$, we obtain dy/dx, which we can write as $f'(x)$. This can in turn be differentiated to give the *second derivative*. This can be written either as d^2y/dx^2 or as $f''(x)$.

Thus if $y = f(x) = ax^n$ then $dy/dx = f'(x) = nax^{(n-1)}$. Also $d^2y/dx^2 = f''(x) = n(n-1)ax^{(n-2)}$.

If $y = ax^3 + bx^2 + cx + e$, then

$$dy/dx = f'(x) = 3ax^2 + 2bx + c$$

$$d^2y/dx^2 = f''(x) = 6ax + 2b$$

$$d^3y/dx^3 = f'''(x) = 6a$$

$$d^4y/dx^4 = f^{IV} = 0.$$

(here $f'''(x)$ is the *third derivative*). Thus for a cubic, the *fourth derivative* is zero.

> **Box 6.1 — Basic Differentials**
>
> If $y = ax^n + bx^{(n-1)} + cx^{(n-2)} + \cdots + px^2 + qx + r$
> Then
> $$dy/dx = nax^{(n-1)} + (n-1)bx^{(n-2)} + (n-2)cx^{(n-3)} + \ldots 2px + q$$
> If $y = uv$ where u and v are both functions of x then
> $$dy/dx = u\, dv/dx + v\, du/dx$$

If dy/dx represents the rate of change of a curve, then d^2y/dx^2 represents the rate of change of the rate of change; put in another context, if y is the distance, dy/dx is the rate of change of distance or the velocity or speed and d^2y/dx^2 is the acceleration.

Consider the function

$$y = 1 + 9x - 6x^2 + x^3$$

$$dy/dx = 9 - 12x + 3x^2 = 3(1 - x)(3 - x)$$

$$d^2y/dx^2 = -12 + 6x$$

$d^2y/dx^2 = 0$ when $x = 2$ and $y = 3$; $dy/dx = 0$ when $x = 1$ and $y = 5$ and when $x = 3$ and $y = 1$. In Table 6.1, we list some values of x and y for this function and have interpolated a curve between the points in Figure 6.3. It can be seen from

Table 6.1 Data for $y = 1 + 9x - 6x^2 + x^3$

x	-3	-2	-1	-0.5	0	0.5	1	1.5	2	2.5	3	3.5	4	4.5	5	5.5	6
y	-107	-49	-15	-5.1	1	4.1	5	4.4	3	1.6	1	1.9	5	11.1	21	35.4	55

Figure 6.3 that there are two points where the slope is horizontal, that is, $dy/dx = 0$ {here (1, 5) and (3, 1)}. These points are known as the *maxima* or *minima* for the curve. A cubic has one maximum and one minimum point — although note that this does not mean a maximum value for all points on the curve, which can come from and go off to infinity. Maxima and minima are turning points where the curve having gone upwards then turns downwards or *vice versa*.

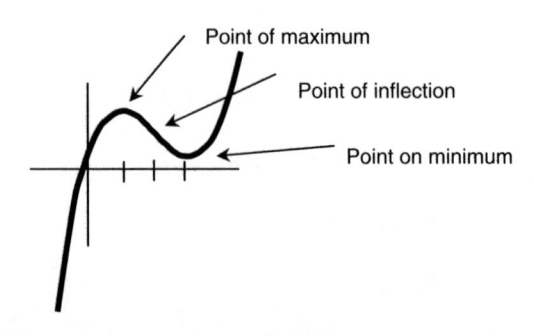

Figure 6.3 A cubic curve.

DIFFERENTIAL AND INTEGRAL CALCULUS

A point where $d^2y/dx^2 = 0$ is known as a *point of inflection*, and a cubic curve has one of them. At a point of inflection, the curve stops bending in one direction and starts bending in the other; in effect, the curve crosses its tangent at that point. In the cubic in Figure 6.3, the curve starting from the bottom left turns clockwise until it reaches the point of inflection where $x = 2$ and $y = 3$ and then starts bending in an anticlockwise direction.

In Chapter 8, we discuss the idea of a radius of curvature at a point on a curve; at a point of inflection, the radius of curvature is infinitely large.

6.2 DIFFERENTIATING TRIGONOMETRIC FUNCTIONS

Before illustrating the use of differentiation, we need to consider the trigonometric functions and how they can be differentiated. We showed in Chapter 5 that

$$\sin(A + B) = \sin A \cos B + \cos A \sin B$$

Hence

$$\sin(\theta + \delta\theta) = \sin\theta \cos\delta\theta + \cos\theta \sin\delta\theta$$

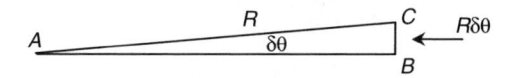

Figure 6.4 Small angles.

In Chapter 5, we defined Sine as side opposite divided by the hypotenuse = BC/AC in Figure 6.4 with cosine as side adjacent over side opposite = AB/AC. As BC becomes smaller and smaller, that is, the angle $\delta\theta$ approaches zero, ultimately $AB = AC$. Thus, cosine $\delta\theta = AB/AC = 1$ as $\delta\theta$ tends to zero. Now, if $AC = R$ and $\delta\theta$ is measured in radians, then $BC = R\delta\theta$. As $\delta\theta$ becomes smaller, BC/AC or $\sin \delta\theta$ becomes $R\delta\theta/R = \delta\theta$. Hence, for small values of $\delta\theta$, $\sin \delta\theta = \delta\theta$. So if $y = \sin\theta$, then

$$y + \delta y = \sin(\theta + \delta\theta)$$

$$= \sin\theta \cos\delta\theta + \cos\theta \sin\delta\theta$$

$$= \sin\theta + \delta\theta \cos\theta$$

Since $y = \sin\theta$, it means that $\delta y = \delta\theta \cos\theta$ or $\delta y/\delta\theta = \cos\theta$. If we revert to the notation that $y = \sin x$ then $dy/dx = \cos x$.

We have also seen in Chapter 5 that

$$\cos(A + B) = \cos A \cos B - \sin A \sin B$$

Hence, if $y = \cos \theta$

$$\text{then } y + \delta y = \cos(\theta + \delta\theta)$$

$$= \cos \theta \cos \delta\theta - \sin \theta \sin \delta\theta$$

$$= \cos \theta - \delta\theta \sin \theta$$

Therefore, $\delta y/\delta\theta = -\sin \theta$. Thus, if $y = \cos x$, then $dy/dx = -\sin x$.

Box 6.2 — Differentiating Trigonometrical Functions (1)

If $y = \sin x$, then $dy/dx = \cos x$ and if $y = \cos x$, then $dy/dx = -\sin x$

Before considering the other trigonometrical functions, we need to consider how to differentiate products. Consider $y = \sin^2 x = \sin x * \sin x$. Let $u = \sin x$ then $du/dx = \cos x$. If $y = \sin^2 x$, then $y = u^2$ and $dy/du = 2u$. Remembering that dy/du means the limit of $\delta y/\delta u$ as δy and δu approach zero, then

$$\frac{\delta y}{\delta x} = \frac{\delta y * \delta u}{\delta x * \delta u} = \frac{\delta y * \delta u}{\delta u * \delta x} = \frac{\delta y}{\delta u} * \frac{\delta u}{\delta x}$$

In the limit

$$(dy/dx) = (dy/du)*(du/dx)$$

$$= 2u*\cos x = 2 \sin x \cos x$$

Thus, if $y = \sin^2 x$, then $dy/dx = 2\sin x \cos x$. Similarly if $y = \sin^3 x$, then $dy/dx = 3 \sin^2 x \cos x$.

We showed above that if $y = u*v$, where u and v are both functions of x, then

$$dy/dx = u\, dv/dx + v\, du/dx$$

We can use this to differentiate $\tan x$. If $y = \tan x = (\sin x)/(\cos x)$, then let $u = \sin x$ for which $du/dx = \cos x$. If $v = 1/(\cos x) = (\cos x)^{-1}$, then

$$dv/dx = -1*(\cos x)^{-2}*(d(\cos x)/dx).$$

$$dv/dx = -1*(\cos x)^{-2}*(-\sin x) = \sin x/(\cos^2 x) = \tan x \sec x$$

(which means that if $y = \sec x$, then $dy/dx = \sec x \tan x$).
So, if $y = \tan x = (\sin x)/(\cos x) = u*v$, then

$$du/dx = \cos x \text{ and } dv/dx = \sec x \tan x$$

DIFFERENTIAL AND INTEGRAL CALCULUS

By substituting these values into "$\{dy/dx = u\, dv/dx + v\, du/dx\}$," we obtain

$$dy/dx = \sin x * \{\sin x/(\cos^2 x)\} + (1/\cos x) * \cos x$$
$$= \sin^2 x/\cos^2 x + 1$$
$$= (\sin^2 x + \cos^2 x)/\cos^2 x = 1/\cos^2 x = \sec^2 x$$

(since $\sin^2 + \cos^2 = 1$). Hence if $y = \tan x$, then $dy/dx = \sec^2 x$. Similar relationships exist for cosec x and cot x. These are summarized in Box 6.3.

Box 6.3 — Differentiating Trigonometrical Functions (2)

If $y = \tan x$ $dy/dx = \sec^2 x$

If $y = \text{cosec } x$ $dy/dx = -\text{cosec } x \cot x$

If $y = \sec x$ $dy/dx = \sec x \tan x$

If $y = \cot x$ $dy/dx = -\text{cosec}^2 x$

6.3 POLYNOMIAL FUNCTIONS

Let us return to the general polynomial of the form

$$y = f(x) = a_0 + a_1 x + a_2 x^2 + a_3 x^3 + a_4 x^4 + \cdots + a_n x^n$$

(Note as an aside that $a_0 = a_0 x^0$ while $a_1 x = a_1 x^1$.) $f(0)$, the value of $f(x)$ when $x = 0$ is such that $f(0) = a_0$.

$$dy/dx = f'(x) = a_1 + 2a_2 x + 3a_3 x^2 + 4a_4 x^3 + \ldots + na_n x^{(n-1)}$$

When x becomes zero, $f'(0) = a_1$ or $a_1 = f'(0)/1$.
If we differentiate $f'(x)$, we obtain $f''(x)$ and so

$$d^2y/dx^2 = f''(x)$$
$$= 2a_2 + 2*3a_3 x + 3*4a_4 x^2 + \cdots + (n-1)na_n x^{(n-2)}$$

Thus, $f''(0) = 2a_2$ or $a_2 = f''(0)/(1*2)$. Again,

$$d^3y/dx^3 = f'''(x)$$
$$= 1*2*3a_3 + 2*3*4a_4 x + \cdots + (n-2)(n-1)\, nx^{(n-3)}$$

Thus, $f'''(0) = 1*2*3*a_3$ or $a_3 = f'''(0)/(1*2*3)$. If we keep doing this, we will find that

$$a_n = f^n(0)/(1*2*3*\cdots *n)$$
$$= f^n(0)/n!$$

where $f^n(0)$ is the value of the nth derivative when $x = 0$. Hence, we can write our original equation as

$$y = f(0) + x^1 f'(0)/1! + x^2 f''(0)/2! + x^3 f'''(0)/3! + \cdots + x^n f^n(0)/n!$$

This is known as Maclaurin's Theorem after the 18th-century Scottish mathematician Colin Maclaurin. Its particular relevance relates to functions such as $f(x) = \sin x$ since $f'(x) = \cos x$, $f''(x) = -\sin x$, $f'''(x) = -\cos x$ and $f^{iv}(x) = \sin x$. Since $\sin(0) = 0$ and $\cos(0) = 1$, then, by measuring x in radians and using Maclaurin's formula we obtain:

$$\sin x = x - x^3/3! + x^5/5! - x^7/7! + \cdots$$

Likewise,

$$\cos x = 1 - x^2/2! + x^4/4! - x^6/6! + \cdots .$$

If we define the exponential function $f(x) = e^x$ as the function for which its differential is itself then $f'(x) = e^x = f''(x)$, etc. Since $e^0 = 1$ (anything to the power of zero = 1), then

$$e^x = 1 + x^1/1! + x^2/2! + x^3/3! + x^4/4! + \cdots$$

$$e^1 = 1 + 1/1! + 1/2! + 1/3! + 1/4! + \cdots$$

$$= 2.7182818 \cdots$$

These series are infinite in that they go on forever but they converge fairly rapidly allowing for the values of functions such as sine, cosine, and e^x to be computed relatively simply. It is not possible to express all functions in the form of a polynomial but many can be expressed this way (Box 6.4).

Box 6.4 — Maclaurin's Theorem

If $y = f(x)$ then

$$y = f(0) + xf'(0)/1! + x^2 f''(0)/2! + x^3 f'''(0)/3! + \cdots + x^n f^n(0)/n!$$

In particular

$$\sin x = x - x^3/3! + x^5/5! - x^7/7! + \cdots$$

$$\cos x = 1 - x^2/2! + x^4/4! - x^6/6! + \cdots$$

$$e^x = 1 + x^1/1! + x^2/2! + x^3/3! + x^4/4! + \cdots$$

Note that in the expansion of e^x given above, we can replace e^x by e^{nx} where n is some constant. The general term in the expansion of e^{nx} will be $(nx)^r/r! = n^r x^r/r!$. Differentiating, we obtain $r*n^r x^{r-1}/r! = n\{n^{r-1} x^{r-1}/(r-1)!\}$ and hence overall, the

DIFFERENTIAL AND INTEGRAL CALCULUS 85

differential of $e^{nx} = n*e^{nx}$. In particular, the differential of $e^{-x} = -e^{-x}$. If $y = e^x$ then $\log_e y = \log_e(e^x) = x \log_e e = x$ (see Box 2.14). Differentiating with respect to y rather than x, we obtain

$$\frac{d(\log_e y)}{dy} = dx/dy.$$

But we defined e^x above as the function for which $dy/dx = e^x = y$. Hence, $dx/dy = 1/y$. Thus

$$\frac{d(\log_e y)}{dy} = 1/y.$$

In general

$$\frac{d(\log_e x)}{dx} = 1/x$$

Also

$$\frac{d(\log_e y)}{dx} = \frac{d(\log_e y)}{dy} * \frac{dy}{dx} = (1/y)*dy/dx$$

In the case of $y = e^{x^2}$, $\log_e y = x^2$ and $(1/y)dy/dx = 2x$. So

$$dy/dx = 2yx = 2xe^{x^2}$$

If we extend all these ideas to three or more dimensions, we will have functions such as $z = f(x, y) = 3x^3 + 2x^2y + y^3 + 4x + 5y$. If we assume that y is treated as a constant, then we can differentiate z with respect to x. Similarly, we can keep x constant and differentiate with respect to y. These are known as the *partial derivatives* and are shown with a curly delta '∂'. Thus, in this example,

$$\partial z/\partial x = 9x^2 + 4xy + 4$$

while

$$\partial z/\partial y = 2x^2 + 3y^2 + 5$$

We shall make use of partial derivatives in Chapter 11.

6.4 BASIC INTEGRATION

The process of differentiation can be reversed by what is known as *integration*.
We have shown that if $y = x^n$ then $dy/dx = nx^{(n-1)}$. Consider the function $f(x) = c + x^{n+1}/(n+1)$ where 'c' is any constant number. If we differentiate this, we obtain $f'(x) = x^n$. The constant 'c' that had been introduced disappears when we differentiate since it is in effect cx^0 and becomes 0. If $y = x^n$, then the function $(c + x^{n+1}/(n+1))$ is said to be the *indefinite integral* of y. It is called indefinite since the number 'c' is unknown at this stage. 'c' is said to be the *constant of integration*.
The integral of y is often shown with an elongated letter 's' of the form $\int$. If $y = bx$, then the integral of y with respect to x,

$$\int y \, dx = \int (bx) \, dx = c + bx^2/2$$

Rather than keep writing "with respect to x", we use the notation from before and write $\int y\, dx$ meaning what happens to y when we integrate with respect to x. If $y = \sin x$, then $\int y\, dx = c - \cos x$ while $\int \cos x\, dx = c + \sin x$. As with basic arithmetic $\int \{f(x) + g(x)\}dx = \int f(x)\, dx + \int (g(x)\, dx$. Also, $\int k f(x)\, dx = k \int f(x)\, dx$ where k is a constant. We have seen that if $y = f(x)*g(x)$, then

$$dy/dx = g(x)*df/dx + f(x)*dg/dx$$

Unfortunately, $\int (f(x)*g(x)\, dx$ cannot be treated in such a neat way. It does not obey this rule.

Given that integration is the reverse of differentiation and given that the integral of $x^n = \int x^n\, dx = c + \{1/(n+1)\}x^{n+1}$, we can apply the same principles if n is negative, with one exception.

Thus, the integral of $x^{-2} = \int x^{-2}\, dx = c - x^{-1}$.

The one exception occurs when $n = -1$. We cannot integrate x^{-1} in this way since $\{1/(n+1)\}$ would become 1/0 and be infinite. We have, however, seen above that

$$\frac{d(\log_e x)}{dx} = 1/x$$

Box 6.5 — Table of Integrals

$$\int x^n\, dx = c + \{1/(n+1)\}\, x^{n+1}$$

Except that $\int x^{-1}\, dx = c + \log_e x$

$$\int \sin x\, dx = c - \cos x$$

$$\int \sin 2x\, dx = c - (1/2)\cos 2x$$

$$\int \cos x\, dx = \sin x + c$$

$$\int e^{nx} = (1/n)e^{nx} + c$$

Also

$$\int \tan x = \log_e(\sec x) + c$$

$$\int \cot x = \log_e(\sin x) + c$$

$$\int \sec x = \log_e(\sec x + \tan x) + c$$

(which can be shown to be $\log_e(\tan (\pi/4 + x/2)) + c)$

$$\int \operatorname{cosec} x = \log_e(\operatorname{cosec} x - \cot x) + c$$

(which can be shown to be $\log_e(\tan x/2) + c)$

DIFFERENTIAL AND INTEGRAL CALCULUS

Knowing this, $\int x^{-1} dx$ becomes $c + \log_e x$. Since the differential of $e^x = e^x$, it follows that $\int e^x dx = e^x + $ constant and $\int e^{nx} dx = (1/n)e^{nx} + $ constant.

For trigonometric functions, the easiest way to show that the integral of tan x, $\int \tan x = \log_e(\sec x)$ is by differentiating the answer. If $v = \sec x$, then

$$dv/dx = \sec x \tan x; \quad d(\log_e v)/dv = 1/v.$$

Thus, $d(\log_e \sec x)/dx = (1/\sec x) * \sec x \tan x = \tan x$.

6.5 AREAS AND VOLUMES

Integration occurs in geomatics in a number of operations, one of which is in determining area. Consider two points on a curve (Figure 6.5) say

$$P(x, y) \text{ and } P'(x + \delta x, y + \delta y)$$

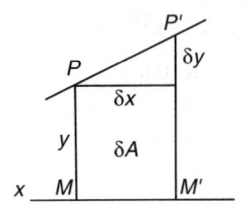

Figure 6.5 Area beneath a curve.

The area of the strip beneath the section of the curve $PP' = MPP'M'$, which we will call δA. Then $\delta A = y\delta x + 1/2 \, \delta y \, \delta x = $ a rectangle plus a triangle. As in previous discussions, as δy and δx become small,

$$\delta A/\delta x = y + 1/2 \, \delta y$$

In the limit,

$$dA/dx = y$$

We can put this another way by saying that if we add or sum all the little bits of δA together as we move along the curve, we shall obtain the whole area under the curve. We can express this summation as $\Sigma \delta A$ where Σ (sigma) means " the sum of."

More particularly as δx tends to zero, we can change to the differential notation and in so doing change the summation symbol from Σ to $\int$ yielding

$$A = \int dA = \int y \, dx$$

Thus, $\int y \, dx$ is the area under a curve. If, for example, $y = x^2$, then the area under the curve $= c + x^3/3$. We still, of course, have the constant of integration that is unknown. If, however, we are interested in the area under the curve between the points where $x = 1$ and 2, then it will be $\{(c + 2^3/3) - (c + 1^3/3)\}$ and the unknown 'c' disappears yielding 7/3.

By setting definite limits to the integration (say from $x = a$ to b), we obtain the *definite integral*. We write this as $\int_a^b y\, dx$, which means "the value of the integral $\int y\, dx$ when x equals b minus the value when x equals a."

If we set limits to the integration, then the result is sometimes written within squared brackets in the form $[I]_{x=a}^{x=b}$ where I is the indefinite integration that is evaluated at $x = b$ and at $x = a$.

So, if $y = x^2$, then $I = x^3/3$ and $\left[\dfrac{x^3}{3}\right]_{x=a}^{x=b} = (b^3 - a^3)/3$.

Returning to Figure 6.4, the area of the sliver in triangle ABC is

$$1/2\ base*height = 1/2\ R*R\delta\theta$$

If we integrate this between 0 and 2π,

$$area = \int_0^{2\pi} 1/2\ R^2\ d\theta = 1/2\ R^2\theta \text{ evaluated within the limits } \theta = 0 \text{ to } \theta = 2\pi.$$

This is

$$\left[\dfrac{R^2\theta}{2}\right]_0^{2\pi} = [\pi R^2 - 0] = \pi R^2.$$

This confirms the well-known formula for the area of a circle.

If $y = ax^2 + bx + c$ and is part of a parabola, then the area between y_1 at x_1, y_2 at x_2 and the x axis will be $[ax^3/3 + bx^2/2 + cx]_{x_1}^{x_2}$

$$area = ax_2^3/3 + bx_2^2/2 + cx_2 - ax_1^3/3 - bx_1^2/2 - cx_1$$
$$= ax_2^3/3 - ax_1^3/3 + bx_2^2/2 - bx_1^2/2 + cx_2 - cx_1$$

Now the lowest common denominator = 6.

Also, $(x_2^3 - x_1^3) = (x_2 - x_1)(x_2^2 + x_2 x_1 + x_1^2)$ and $(x_2^2 - x_1^2) = (x_2 - x_1)(x_2 + x_1)$.

$$area = (1/6)(x_2 - x_1)\{2ax_2^2 + 2ax_2 x_1 + 2ax_1^2 + 3bx_2 + 3bx_1 + 6c\}$$
$$= (1/6)(x_2 - x_1)\{ax_2^2 + bx_2 + c + ax_1^2 + bx_1 + c$$
$$+ ax_2^2 + 2ax_2 x_1 + ax_1^2 + 2bx_2 + 2bx_1 + 4c\}$$
$$= (1/6)(x_2 - x_1)\{y_2 + y_1 + a(x_2 + x_1)^2 + 2b(x_2 + x_1) + 4c\}$$
$$= (1/6)(x_2 - x_1)\{y_2 + y_1 + 4a^2\left(\dfrac{x_2 + x_1}{2}\right)^2 + 4b\left(\dfrac{x_2 + x_1}{2}\right) + 4c\}$$
$$= (1/6)(x_2 - x_1)\{y_2 + y_1 + 4y_m\}$$

where y_m is the value of y at the point midway between x_1 and x_2 where x has the value $(x_1 + x_2)/2$.

The formula, known as Simpson's Rule after the 18th-century mathematician Thomas Simpson, provides a method to determine areas of irregular shapes drawn on a map. Thus, in Figure 6.6, we can divide an area into a series of strips of equal width and measure the length of each straight line. If we assume that each strip is of width 'w', then if we treat each strip as a parallelogram; its area is $w*(Y_r + Y_{r+1})/2$.

DIFFERENTIAL AND INTEGRAL CALCULUS 89

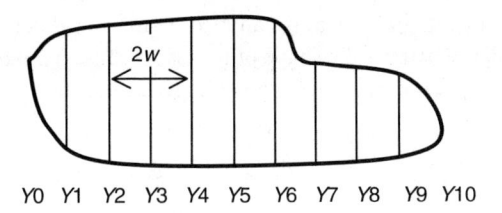

Figure 6.6 Area of an irregular shape.

For all of Figure 6.6,

$$area = (w/2)*(Y_0 + 2Y_1 + 2Y_2 + 2Y_3 + 2Y_4 + 2Y_5 + 2Y_6 + 2Y_7 + 2Y_8 + 2Y_9 + Y_{10})$$

This assumes that the boundaries are all straight lines rather than curves. Using Simpson's Rule, we would firstly need to make sure that there is an even number of strips and then treat each pair of strips. The area of the first pair, for example, is:

$$\text{Area of double strip} = (1/6)*2w*(Y_0 + 4Y_1 + Y_2).$$

The area for the whole shape would then be

$$area = (w/3)\{Y_0 + 4Y_1 + 2Y_2 + 4Y_3 + 2Y_4 + 4Y_5 + 2Y_6 + 4Y_7 + 2Y_8 + 4Y_9 + Y_{10}\}$$

This will be a better approximation as it assumes that the boundaries are curves that approximate to sections of a parabola.

Box 6.6 — Area of Irregular Shape

For strips of width 'w' and lengths Y_i ($i = 0$ to n) then
By Trapezium Rule

$$area = (w/2)*(Y_0 + 2Y_1 + 2Y_2 + 2Y_3 + 2Y_4 + 2Y_5 + \cdots + 2Y_{n-2} + 2Y_{n-1} + Y_n)$$

By Simpson's Rule:

$$area = (w/3)\{Y_0 + 4Y_1 + 2Y_2 + 4Y_3 + 2Y_4 + 4Y_5 + \cdots + 2Y_{n-2} + 4Y_{n-1} + Y_n\}$$

where n is an even number.

The principles of integral calculus can also be used to determine volumes. The volume of a cylinder, whatever the shape of its base = base area∗height provided the sides are parallel. Thus, the volume of the cylinder in Figure 6.7 = Ah.

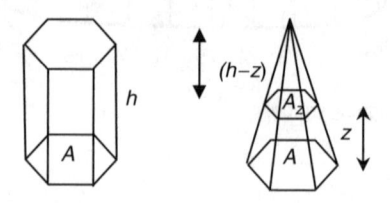

Figure 6.7 Volumes of a cylinder and cone.

For the cone (be it a pyramid or any other base shape with a base area A and height h), the area A_z of cross-sectional slice at height z above the base will be in proportion to the base A but reduced both in terms of length and breadth to an amount $\{(h-z)/h\}$. Its area will therefore be $A_z = \{(h-z)/h\} * \{(h-z)/h\} * A = \{A/h^2\}\{h^2 - 2hz + z^2\}$.
If the cross-section is δz thick, then its volume will be

$$\delta V = (A/h^2)(h^2 - 2hz + z^2)\delta z$$

If we sum all this from the base to the apex

$$V = (A/h^2) \int_{z=0}^{z=h} (h^2 - 2hz + z^2) \, dz = (A/h^2) \left[h^2 z - hz^2 + z^3/3 \right]_{z=0}^{z=h}$$

$$= (1/3) \, Ah^3/h^2 = (1/3)Ah = \text{one-third of base area}*\text{height}$$

Box 6.7 — Volume of a Cone or Pyramid

The volume of a pyramid or cone is one-third of its base area times its height.

CHAPTER 7

Matrices, Determinants, and Vectors

CONTENTS

7.1 Basic Matrix Operations 91
7.2 Determinants 94
7.3 Related Matrices 95
7.4 Applying Matrices 98
7.5 Rotations and Translations 99
7.6 Simplifying Matrices 105
7.7 Vectors 109

7.1 BASIC MATRIX OPERATIONS

Matrices are a mathematical form of shorthand. At its simplest level, a *matrix* is a set of numbers aligned in rows and columns in the form of a rectangle and then enclosed in brackets. Each number within the matrix is an *element* and the whole set of numbers may be referred to as an *array*. Rather than discussing each of the elements within the array, we can simply refer to the matrix as a whole and call it **M**.

For example,

$$\mathbf{M} = \begin{pmatrix} 1 & 2 & 3 \\ 4 & 5 & 6 \end{pmatrix}$$

is a matrix with the first six integers arranged in two rows and three columns and is called a 2∗3 matrix.

$$\begin{pmatrix} 7 \\ 8 \\ 9 \end{pmatrix}$$

is a column matrix with three rows and one column or 3∗1. If the number of rows equals the number of columns, then the matrix is said to be *square*.

$$\begin{pmatrix} -1 & 0 & 0 \\ 0 & 2 & 0 \\ 0 & 0 & -3 \end{pmatrix}$$

is a 3∗3 square matrix and since in this case the only numbers other than zero lie along the diagonal, it is called a *diagonal matrix*. The diagonal from top left to bottom right is called the leading diagonal. If all the numbers in the leading diagonal are 1 and all the other elements are zero, then the matrix is called an *identity matrix*.

$$\begin{pmatrix} 1 & 0 \\ 0 & 1 \end{pmatrix} \text{ and } \begin{pmatrix} 1 & 0 & 0 \\ 0 & 1 & 0 \\ 0 & 0 & 1 \end{pmatrix}$$

are both identity matrices and are often written as **I**. The way in which matrices are manipulated follows certain rules that are ideally suited to handling in a computer, as the operations are, in general, repetitive.

Two matrices can be added or subtracted if they are the same size (Box 7.1). This is done by adding or subtracting the corresponding elements. Thus,

$$\begin{pmatrix} 1 & 2 & 3 & 4 \\ 5 & 6 & 7 & 8 \end{pmatrix}$$

is a 2∗4 matrix. If we add another 2∗4 matrix

$$\begin{pmatrix} 6 & -5 & 4 & -3 \\ 9 & -8 & 7 & 6 \end{pmatrix}$$

we obtain

$$\begin{pmatrix} 1+6 & 2-5 & 3+4 & 4-3 \\ 5+9 & 6-8 & 7+7 & 8+6 \end{pmatrix} = \begin{pmatrix} 7 & -3 & 7 & 1 \\ 14 & -2 & 14 & 14 \end{pmatrix}$$

The matrix

$$\begin{pmatrix} 1 & 2 & 3 \\ 4 & 5 & 6 \end{pmatrix}$$

cannot, however, be added to or subtracted from

$$\begin{pmatrix} 7 \\ 8 \\ 9 \end{pmatrix}$$

because the two matrices are not of the same size. If the entire matrix is multiplied by a number, then each element must be multiplied by that number. Thus, if

$$\mathbf{M} = \begin{pmatrix} 1 & 2 & 3 & 4 \\ 5 & 6 & 7 & 8 \end{pmatrix}, \text{ then } 4\mathbf{M} = \begin{pmatrix} 4 & 8 & 12 & 16 \\ 20 & 24 & 28 & 32 \end{pmatrix}$$

Note that to refer to a matrix, we write it in **bold** lettering (e.g., **M** above).

MATRICES, DETERMINANTS, AND VECTORS

> **Box 7.1 — Matrix Addition and Subtraction**
>
> If **A** is a matrix, the number in the ith row and jth column $= a_{ij}$ and **B** is the same size with the ith row and jth column b_{ij}, then the sum of $\mathbf{A} + \mathbf{B} = \mathbf{C}$ where $c_{ij} = a_{ij} + b_{ij}$. Likewise, if $\mathbf{A} - \mathbf{B} = \mathbf{D}$, then $d_{ij} = a_{ij} - b_{ij}$

To multiply one matrix by another the number of columns in the first matrix must equal the number of rows in the second one. This is because the multiplication is carried out in a special way, which creates a new matrix with the same number of rows as the first and the same number of columns as the second. In other words, if the first matrix has 'p' rows and 'q' columns and the second has 'q' rows and 'r' columns, then the product of the two will have 'p' rows and 'r' columns.

Consider a 2∗2 matrix

$$\mathbf{A} = \begin{pmatrix} a & b \\ d & e \end{pmatrix}$$

with numbers a, b, d, and e. Let **B** be a 2∗1 matrix with numbers u and w such that

$$\mathbf{B} = \begin{pmatrix} u \\ w \end{pmatrix}$$

Then

$$\mathbf{A} * \mathbf{B} = \begin{pmatrix} a & b \\ d & e \end{pmatrix} * \begin{pmatrix} u \\ w \end{pmatrix} = \begin{pmatrix} au + bw \\ du + ew \end{pmatrix}$$

For the first row, multiply each element across the row by the element that is in the same position down the column of the second matrix. Repeat for the second row of the first matrix and the first column of the second. The result of multiplying a 2∗2 matrix times a 2∗1 matrix is a matrix with 2 rows and 1 column. Conversely, we cannot multiply a 2∗1 matrix times a 2∗2, since the columns in the first matrix are not as many as the rows in the second one. In this case, **B**∗**A** does not exist.

The same procedure is followed with a 3∗3 multiplying a 3∗2:

$$\begin{pmatrix} a & b & c \\ d & e & f \\ g & h & i \end{pmatrix} * \begin{pmatrix} u & x \\ v & y \\ w & z \end{pmatrix} = \begin{pmatrix} au + bv + cw & ax + by + cz \\ du + ev + fw & dx + ey + fz \\ gu + hv + iw & gx + hy + iz \end{pmatrix} = \mathbf{C}$$

which is a 3∗2 matrix.

In the above example, if in **A** the numbers $a = e = i = 1$ and all the other elements of **A** are zero, that is, if **A** is the identity matrix

$$\mathbf{I} = \begin{pmatrix} 1 & 0 & 0 \\ 0 & 1 & 0 \\ 0 & 0 & 1 \end{pmatrix} \text{ and if } \mathbf{B} = \begin{pmatrix} u & x \\ v & y \\ w & z \end{pmatrix}$$

then when we multiply **A**∗**B** to obtain **C**, all the elements of **C** will be the same as those in **B**, hence

$$\mathbf{I} * \mathbf{B} = \mathbf{B}$$

Once again, if **A** is the 3∗3 matrix and **B** is the 3∗2, we can multiply **A** times **B** to obtain a new matrix **C** but we cannot multiply **B** times **A**, since the number of columns in the first does not equal the number of rows in the second. If, however, the number of rows in **A** equals the number of columns in **B**, then we could obtain both **A**∗**B** and **B**∗**A**. Thus, if **A** is 4∗2 and **B** is 2∗4, then **A**∗**B** is 4∗4 but **B**∗**A** is 2∗2 and **A**∗**B** ≠ **B**∗**A**. Thus, unlike ordinary arithmetic, the order of multiplication is of critical importance (Box 7.2).

Box 7.2 — Matrix Multiplication

If **A** is a matrix with the element in the *i*th row and *j*th column a_{ij} and **B** has the same number of rows as **A** has columns and has in its *i*th row and *j*th column b_{ij}, then **A**∗**B** = **C** where c_{ij} = the sum of the elements of row '*i*' in **A** multiplied by the corresponding elements of column '*j*' in **B**, i.e. $\{a_{i1}b_{1j} + a_{i2}b_{2j} + a_{i3}b_{3j} + \cdots + a_{in}b_{nj}\}$

There are, however, exceptions to the rule that **A**∗**B** ≠ **B**∗**A**, for example, when **B** is such that **A**∗**B** = **I** (the identity matrix). Consider the special case where

$$\mathbf{A} = \begin{pmatrix} a & b \\ d & e \end{pmatrix}$$

and

$$\mathbf{B} = \frac{1}{ae - bd} * \begin{pmatrix} e & -b \\ -d & a \end{pmatrix}$$

If we multiply **A** and **B** (=**A**∗**B**) we obtain

$$\begin{pmatrix} 1 & 0 \\ 0 & 1 \end{pmatrix}$$

the identity matrix. We obtain the same result if we multiply **B** and **A** (=**B**∗**A**). The matrix **B** that has this special property is said to be the *inverse* of **A** and is written as $\mathbf{A}^{-1}$. It has the property that $\mathbf{A}\mathbf{A}^{-1} = \mathbf{A}^{-1}\mathbf{A} = \mathbf{I}$.

7.2 DETERMINANTS

The number ($ae - bd$) that was used to scale the whole multiplication to make the diagonal numbers all equal to 1 is a special number for the matrix **A**, which is called its *determinant*. This is usually written with straight-line brackets, so that

$$|\mathbf{A}| = \begin{vmatrix} a & b \\ d & e \end{vmatrix} = ae - bd$$

If it happens that $ae = bd$, then $1/|\mathbf{A}| = 1/0$, which is infinite; hence, the matrix **A** has no inverse. In this case, the matrix is said to be *singular*.

For a 3∗3 matrix, the process is slightly more complicated. The calculation is broken down into 3 stages where

MATRICES, DETERMINANTS, AND VECTORS

$$\begin{vmatrix} a & b & c \\ d & e & f \\ g & h & i \end{vmatrix} = 3 \text{ parts} \quad \begin{vmatrix} a & . & . \\ . & e & f \\ . & h & i \end{vmatrix} \text{ and } \begin{vmatrix} . & b & . \\ d & . & f \\ g & . & i \end{vmatrix} \text{ and } \begin{vmatrix} . & . & c \\ d & e & . \\ g & h & . \end{vmatrix}$$

Each of these subcomponents contains a 2*2 determinant that is known as the *minor* of the element in the first row. The value of the 3*3 determinant is then defined as (Box 7.3)

$$|\mathbf{A}| = a(ei - fh) - b(di - fg) + c(dh - eg)$$

Box 7.3 — The Determinant of A

If **A** is a 3*3 square matrix such that

$$|\mathbf{A}| = \begin{vmatrix} a & b & c \\ d & e & f \\ g & h & i \end{vmatrix}, \text{ then } |\mathbf{A}| = a * \begin{vmatrix} e & f \\ h & i \end{vmatrix} - b * \begin{vmatrix} d & f \\ g & i \end{vmatrix} + c * \begin{vmatrix} d & e \\ g & h \end{vmatrix}$$

Thus, the determinant of $\mathbf{A} = |\mathbf{A}| = a(ei - fh) - b(di - fg) + c(dh - eg)$. If $|\mathbf{A}| = 0$, the determinant is singular. For a 4*4 square matrix the determinant becomes

$$\begin{vmatrix} a & b & c & p \\ d & e & f & q \\ g & h & i & r \\ j & k & l & s \end{vmatrix} = a * \begin{vmatrix} e & f & q \\ h & i & r \\ k & l & s \end{vmatrix} - b * \begin{vmatrix} d & f & q \\ g & i & r \\ j & l & s \end{vmatrix} + c * \begin{vmatrix} d & e & q \\ g & h & r \\ j & k & s \end{vmatrix} - p * \begin{vmatrix} d & e & f \\ g & h & i \\ j & k & l \end{vmatrix}$$

where each of the 3*3 minor determinants has to be evaluated as above. Note the overall symmetry of the operation and the alternating signs + − + − + −, etc.

7.3 RELATED MATRICES

Before considering the importance of matrices in geomatics, we need to consider two other concepts. The first is the idea of the *transpose* of a matrix **A**, which is denoted $\mathbf{A}^T$. $\mathbf{A}^T$ is formed by changing the rows in **A** into columns in $\mathbf{A}^T$ and the columns in **A** into rows in $\mathbf{A}^T$. Thus, if

$$\mathbf{A} = \begin{pmatrix} a & b & c & d \\ e & f & g & h \end{pmatrix}, \text{ then } \mathbf{A}^T = \begin{pmatrix} a & e \\ b & f \\ c & g \\ d & h \end{pmatrix}$$

If **A** has '*m*' rows and '*n*' columns, then $\mathbf{A}^T$ will have '*n*' rows and '*m*' columns and we can multiply **A** and $\mathbf{A}^T$ ($=\mathbf{A}*\mathbf{A}^T$) and also $\mathbf{A}^T$ and $\mathbf{A}(=\mathbf{A}^T*\mathbf{A})$ (Box 7.4).

In the particular case where $\mathbf{A} = (x\ y\ z)$, then $\mathbf{A}^T = \begin{pmatrix} x \\ y \\ z \end{pmatrix}$ and

$$\mathbf{A} * \mathbf{A}^T = (x^2 + y^2 + z^2),\ \text{which is a}\ 1*1\ \text{matrix}$$

On the other hand, if we multiply A^T and $A (= \mathbf{A}^T * \mathbf{A}) = \begin{pmatrix} x \\ y \\ z \end{pmatrix} (x\ y\ z)$, we obtain

$$\mathbf{A}^T * \mathbf{A} = \begin{pmatrix} xx & xy & xz \\ yx & yy & yz \\ zx & zy & zz \end{pmatrix},\ \text{which is a}\ 3*3\ \text{matrix}$$

Box 7.4 — Matrix Multiplication (2)

In general (although not in all cases)

$$\mathbf{A} * \mathbf{B} \neq \mathbf{B} * \mathbf{A}\ \text{and}\ \mathbf{A}\mathbf{A}^T \neq \mathbf{A}^T\mathbf{A}$$

The order in which matrices are multiplied together is of crucial importance.

The other idea we need to introduce is to label every element according to its row and column. Rather than saying $\mathbf{A} = (a\ b\ c\ d)$ we can say that $\mathbf{A} = (a_{11}\ a_{12}\ a_{13}\ a_{14})$.

For a 4*4 matrix,

$$\mathbf{A} = \begin{pmatrix} a_{11} & a_{12} & a_{13} & a_{14} \\ a_{21} & a_{22} & a_{23} & a_{24} \\ a_{31} & a_{32} & a_{33} & a_{34} \\ a_{41} & a_{42} & a_{43} & a_{44} \end{pmatrix}$$

In general, a_{ij} is the element in the ith row and jth column. In the transpose $\mathbf{A}^T$ it becomes the jth row of the ith column or a_{ji}.

We can think of a square matrix $\mathbf{A}$ as having elements occupying the squares in a chequer board (shown in Figure 7.1 as 8*8; but we will assume that it is $n*n$ where n is a positive integer). The board would contain an element such as a_{ij} for the number in the 'i'th row and 'j'th column. If we strip out the 'i'th row and 'j'th column as in Figure 7.1(b), then we are left with $(n-1)*(n-1)$ matrix that is called the *submatrix of* a_{ij}. The determinant of this submatrix of a_{ij} is known as the *minor of* a_{ij}.

If we assign the minor of a_{ij} the name M_{ij}, then just as we saw in Box 7.3, we need to alternate the signs (plus then minus then plus then minus then plus, etc.). We can think of black squares as plus and white squares as minus or we can say that the sign is $(-1)^{(i-j)}$. We now define the *cofactor* of the element a_{ij} in A as $M_{ij}(-1)^{(i-j)}$ and write it as A_{ij}. A cofactor is a determinant and reduces to a single number.

MATRICES, DETERMINANTS, AND VECTORS

(a) (b)

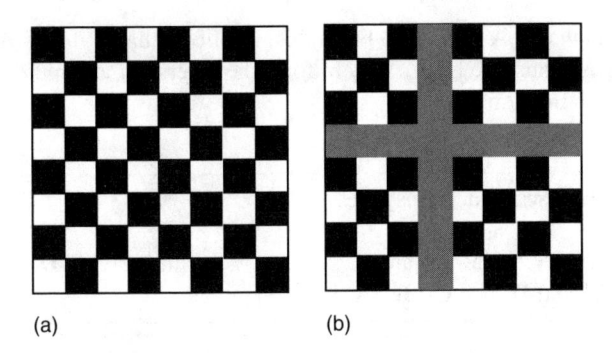

Figure 7.1 Submatrices and minors.

The overall value of the determinant of the matrix **A** is obtained from any row and its related cofactors in the form

$$|A| = a_{11}A_{11} + a_{12}A_{12} + a_{13}A_{13} + \cdots + a_{1n}A_{1n}$$

Alternatively,

$$|A| = a_{i1}A_{i1} + a_{i2}A_{i2} + a_{i3}A_{i3} + \cdots + a_{in}A_{in}$$

basing the calculation on row 'i'.

Thus, an $n*n$ matrix has a determinant that is the sum of n products between the row elements and their respective cofactors. Each cofactor can be evaluated as the sum of its minors and so on down to the elements of a determinant that is $2*2$. This is an extension of what we showed in Box 7.3.

We now define as the *adjugate* matrix $\mathbf{A}^*$ of a square matrix **A** a new matrix formed by cofactors of the elements of **A** in the transposed position. Thus, for a $4*4$ matrix, the terms are

$$\mathbf{A}^* = \begin{pmatrix} A_{11} & A_{21} & A_{31} & A_{41} \\ A_{12} & A_{22} & A_{32} & A_{42} \\ A_{13} & A_{23} & A_{33} & A_{43} \\ A_{14} & A_{24} & A_{34} & A_{44} \end{pmatrix}$$

If

$$\mathbf{A} = \begin{pmatrix} a_{11} & a_{12} & a_{13} & a_{14} \\ a_{21} & a_{22} & a_{23} & a_{24} \\ a_{31} & a_{32} & a_{33} & a_{34} \\ a_{41} & a_{42} & a_{43} & a_{44} \end{pmatrix}$$

then we can show that

$$\mathbf{AA}^* = \begin{pmatrix} |A| & 0 & 0 & 0 \\ 0 & |A| & 0 & 0 \\ 0 & 0 & |A| & 0 \\ 0 & 0 & 0 & |A| \end{pmatrix} = |A|\mathbf{I}$$

We can also show that $\mathbf{A}^*\mathbf{A} = |\mathbf{A}|\mathbf{I}$. From this, it follows that $\mathbf{A}^*/|\mathbf{A}| = \mathbf{A}^{-1}$, which is the inverse of $\mathbf{A}$. Thus, we have a way to form the inverse of any matrix by computing its adjugate (Box 7.5).

Box 7.5 — Inverses and Transposes

If $\mathbf{C} = \mathbf{AB}$, then $\mathbf{C}^\mathrm{T} = \mathbf{B}^\mathrm{T}\mathbf{A}^\mathrm{T}$ and $\mathbf{C}^{-1} = \mathbf{B}^{-1}\mathbf{A}^{-1}$. Similarly, if $\mathbf{D} = \mathbf{ABC}$, then $\mathbf{D}^\mathrm{T} = \mathbf{C}^\mathrm{T}\mathbf{B}^\mathrm{T}\mathbf{A}^\mathrm{T}$ and $\mathbf{D}^{-1} = \mathbf{C}^{-1}\mathbf{B}^{-1}\mathbf{A}^{-1}$

7.4 APPLYING MATRICES

Let us now consider some simple applications of matrices, recalling that matrices are a mathematical form of shorthand and only actually come into their own when dealing with large sets of numbers. Matrices are made up of numbers such as a_{ij} and a_{kl}, which will normally be subject to the four simple operations of addition, subtraction, multiplication, and division. The answers to their manipulation are then put in the appropriate location in the new resulting matrix.

The processes of manipulating matrices are ideally suited to handling by computer as the operations are sequential and routine. Consider a very simple example that involves the intersection of two straight lines illustrated in Box 7.6. The calculation involves the conversion of a square matrix into its inverse ($\mathbf{A}$ into $\mathbf{A}^{-1}$) and the multiplication of two matrices ($\mathbf{A}^{-1}*\mathbf{B}$).

In the example in Box 7.6, if the determinant $|\mathbf{A}|$ = zero (which happens when $a_{11}a_{22} = a_{12}a_{21}$), there is no solution. This will happen when the two lines are parallel and therefore, do not intersect in finite space.

Box 7.6 — The Intersection of Two Lines

In Chapter 3, we calculated the intersection of two straight lines.
Let us express these in the form

$$a_{11}x + a_{12}y = b_{11}$$

$$a_{21}x + a_{22}y = b_{21}$$

or

$$\begin{pmatrix} a_{11} & a_{12} \\ a_{21} & a_{22} \end{pmatrix} \begin{pmatrix} x \\ y \end{pmatrix} = \begin{pmatrix} b_{11} \\ b_{21} \end{pmatrix} \quad \text{or} \quad \mathbf{AX} = \mathbf{B}$$

Multiply both sides of the expression by $\mathbf{A}^{-1}$ to give

$$\mathbf{A}^{-1}\mathbf{AX} = \mathbf{A}^{-1}\mathbf{B}$$

—Continued

MATRICES, DETERMINANTS, AND VECTORS

Box 7.6 — The Intersection of Two Lines (Continued)

Since $\mathbf{A}^{-1}\mathbf{A} = \mathbf{I}$, we have

$$\mathbf{X} = \mathbf{A}^{-1}\mathbf{B}$$

As seen previously,

$$\mathbf{A}^{-1} = \{1/(a_{11}a_{22} - a_{12}a_{21})\} * \begin{pmatrix} a_{22} & -a_{12} \\ -a_{21} & a_{11} \end{pmatrix}$$

Hence,

$$\mathbf{X} = \begin{pmatrix} x \\ y \end{pmatrix} = \{1/(a_{11}a_{22} - a_{12}a_{21})\} * \begin{pmatrix} a_{22} & -a_{12} \\ -a_{21} & a_{11} \end{pmatrix} \begin{pmatrix} b_{11} \\ b_{21} \end{pmatrix}$$

or

$$\begin{pmatrix} x \\ y \end{pmatrix} = \{1/(a_{11}a_{22} - a_{12}a_{21})\} * \begin{pmatrix} a_{22}b_{11} - a_{12}b_{21} \\ -a_{21}b_{11} + a_{11}b_{21} \end{pmatrix}$$

or

$$x = (a_{22}b_{11} - a_{12}b_{21})/(a_{11}a_{22} - a_{12}a_{21})$$

$$y = (a_{11}b_{21} - a_{21}b_{11})/(a_{11}a_{22} - a_{12}a_{21})$$

7.5 ROTATIONS AND TRANSLATIONS

Matrix algebra is very useful when dealing with coordinate systems and changes in the position and orientation of the basic axes of reference. Let us first deal with two dimensions and let the origin be changed by an amount (a, b) from origin 1 to origin 2 (see Figure 7.2).

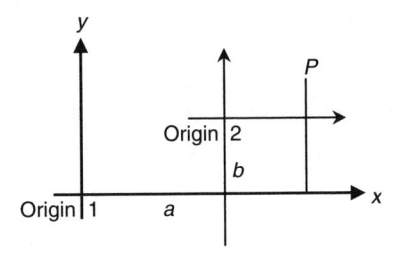

Figure 7.2 Shift or translation of origin.

If the coordinates of any point P are (x_1, y_1) when referred to the old origin 1, then relative to the new origin 2, they are $(x_1 - a, y_1 - b)$.

For a series of points: $\mathbf{X}_n = \mathbf{X}_o - \mathbf{A}$, where '$n$' means new, '$o$' means old and the matrix

$$\mathbf{X} = \begin{pmatrix} x_i \\ y_i \end{pmatrix} \text{ and } \mathbf{A} = \begin{pmatrix} a \\ b \end{pmatrix}$$

This relationship, $\mathbf{X}_n = \mathbf{X}_o - \mathbf{A}$ (where $\mathbf{X}_n$ is the new $\mathbf{X}$ and $\mathbf{X}_o$ the old), will apply in any number of dimensions, especially three. It is often referred to as *translation* meaning moving the origin from one point to another.

If we want to change the scale of our projection by a factor 's,' then $\mathbf{X}_n = s\mathbf{X}$, that is, each of the values of x and y will be multiplied by 's.' If we want to change the scale in the x-direction by a different amount than in the y direction, then we need to multiply by

$$\mathbf{S} = \begin{pmatrix} s_1 & 0 \\ 0 & s_2 \end{pmatrix} \text{ so that } \mathbf{X}_n = \mathbf{SX}$$

If we then change the origin, we obtain $\mathbf{X}_n = \mathbf{SX} - \mathbf{A}$. If we want to change the origin and then change the scale, then $\mathbf{X}_n = \mathbf{S}(\mathbf{X} - \mathbf{A})$. If we want to rotate the axes about their origin through an angle θ, then matrix algebra provides a convenient way to describe this (Figure 7.3). Consider a point P and a set of axes $OX_o\ OY_o$. The coordinates of P are given by $x_o = OA$ and $y_o = AP$. If the axes are rotated clockwise through an angle θ to $OX_n\ OY_n$, then the new coordinates will be $x_n = OB$ and $y_n = BP$. Let angle $POA = \emptyset$ so that $PA = y_o = OP \sin \emptyset$ and $OA = x_o = OP \cos \emptyset$.

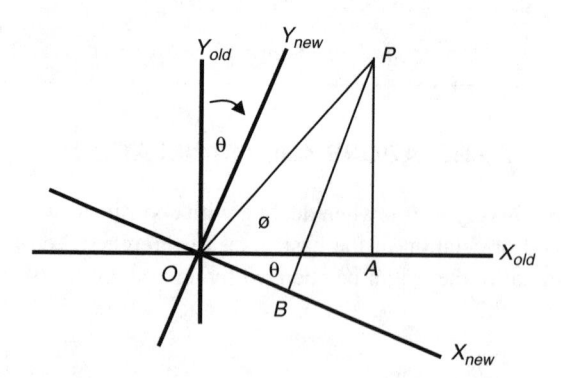

Figure 7.3 Rotation of axes.

$$BP = y_n = OP \sin(\theta + \emptyset) = OP(\sin \theta \cos \emptyset + \cos \theta \sin \emptyset)$$
$$= OP \sin \theta \cos \emptyset + OP \cos \theta \sin \emptyset$$
$$y_n = y_o \cos \theta + x_o \sin \theta$$

Also

$$OB = x_n = OP \cos(\theta + \emptyset) = OP(\cos \theta \cos \emptyset - \sin \theta \sin \emptyset)$$
$$x_n = x_o \cos \theta - y_o \sin \theta$$

Put in matrix form

$$\begin{pmatrix} x_n \\ y_n \end{pmatrix} = \begin{pmatrix} \cos \theta & -\sin \theta \\ \sin \theta & \cos \theta \end{pmatrix} \begin{pmatrix} x_o \\ y_o \end{pmatrix} \text{ or } \mathbf{X}_n = \mathbf{RX}_o$$

MATRICES, DETERMINANTS, AND VECTORS

where

$$\mathbf{R} = \begin{pmatrix} \cos\theta & -\sin\theta \\ \sin\theta & \cos\theta \end{pmatrix}$$

Note that the determinant of $\mathbf{R} = \cos^2\theta + \sin^2\theta = 1$. The inverse of $\mathbf{R}$ is

$$\mathbf{R}^{-1} = \begin{pmatrix} \cos\theta & \sin\theta \\ -\sin\theta & \cos\theta \end{pmatrix}$$

Hence,

$$\mathbf{R}\mathbf{R}^{-1} = \begin{pmatrix} \cos^2\theta + \sin^2\theta & \sin\theta\cos\theta - \sin\theta\cos\theta \\ \sin\theta\cos\theta - \sin\theta\cos\theta & \sin^2\theta + \cos^2\theta \end{pmatrix} = \begin{pmatrix} 1 & 0 \\ 0 & 1 \end{pmatrix}$$

This is the identity matrix **I**. Any matrix that when multiplied by its inverse gives the identity matrix is called an *orthogonal matrix*.

The processes of rotating axes but retaining their rectangularity, moving the origin and perhaps applying a uniform scale change, is called a *similarity transformation*. It is often used in photogrammetry and in computer-generated models. We shall consider an example in Chapter 9. The above derivation assumes that a *positive rotation is clockwise* as we look down on the plan (Box 7.7). Since $\cos(-\theta) = \cos\theta$ and $\sin(-\theta) = -\sin\theta$, if we were to regard an anticlockwise rotation as positive, we would have

$$\begin{pmatrix} x_n \\ y_n \end{pmatrix} = \begin{pmatrix} \cos\theta & \sin\theta \\ -\sin\theta & \cos\theta \end{pmatrix} \begin{pmatrix} x_o \\ y_o \end{pmatrix}$$

Box 7.7 — Rotation of Axes

To transform old coordinates to new coordinates when rotating the x and y orthogonal axes through an angle θ measured as positive in a clockwise direction:

$$\begin{pmatrix} x_n \\ y_n \end{pmatrix} = \begin{pmatrix} \cos\theta & -\sin\theta \\ \sin\theta & \cos\theta \end{pmatrix} \begin{pmatrix} x_o \\ y_o \end{pmatrix}$$

In general, when changing the direction of the axes, $\mathbf{X}_n = \mathbf{R}\mathbf{X}_o$ where $\mathbf{R}$ is called the *rotation matrix*. Because in geomatics we are often talking about bearings that are measured in a clockwise direction, we will use the convention that rotations are positive if clockwise. In photogrammetry, however, the standard convention uses the opposite, which inevitably leads to confusion. The crucial point is "the sign of the sine." The problem arises because what is clockwise when looking upward is anticlockwise when looking downward. An example of a photogrammetric calculation is given later.

$\mathbf{R}$ as defined above is an orthogonal matrix; but there are occasions when a rectangular system needs to be skewed. For example, in Figure 7.4, the old grid that is rectangular has been skewed to form a new grid for ease of representation on a flat piece of paper. The axes are no longer at right angles, although the distances parallel to the corresponding x- and y-axis remain the same ($AP = AP'$).

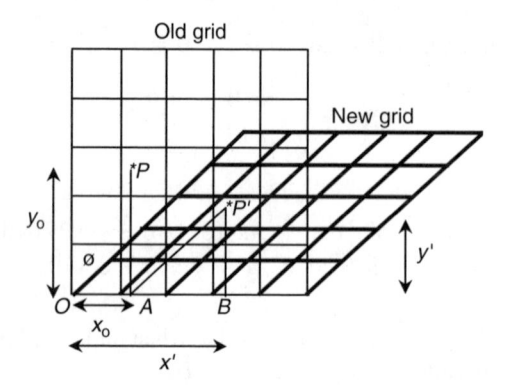

Figure 7.4 A skewed grid.

Thus, in the old grid, the coordinates of P are (x_o, y_o) or (OA, AP), while the distances for point P' on the new grid are (OA, AP') unless there is any change in scale. However, relative to the old grid, the coordinates of P' have become (x', y') or (OB, BP'). $x' = OB = OA + AP' \sin ø = x + y \sin ø$, where $ø$ is the clockwise angle by which the y-axis has been rotated.

$$y' = BP' = AP' \cos ø = y \cos ø$$

Thus

$$\begin{pmatrix} x' \\ y' \end{pmatrix} = \begin{pmatrix} 1 & \sin ø \\ 0 & \cos ø \end{pmatrix} \begin{pmatrix} x \\ y \end{pmatrix} \quad \text{or} \quad \mathbf{X'} = \mathbf{MX} \quad \text{where } \mathbf{M} = \begin{pmatrix} 1 & \sin ø \\ 0 & \cos ø \end{pmatrix}$$

Here **M** is NOT an orthogonal transformation.

If we want to reduce the scale in the 'y' direction by a factor 's_y' so that 'y' would become '$s_y y$' while in the 'x' direction we applied a scale factor of 's_x' so that 'x' becomes '$s_x x$', then

$$\begin{pmatrix} x' \\ y' \end{pmatrix} = \begin{pmatrix} 1 & \sin ø \\ 0 & \cos ø \end{pmatrix} \begin{pmatrix} s_x & 0 \\ 0 & s_y \end{pmatrix} \begin{pmatrix} x \\ y \end{pmatrix} = \begin{pmatrix} s_x & s_y \sin ø \\ 0 & s_y \cos ø \end{pmatrix} \begin{pmatrix} x \\ y \end{pmatrix}$$

Note that

$$\begin{pmatrix} 1 & \sin ø \\ 0 & \cos ø \end{pmatrix} \begin{pmatrix} s_x & 0 \\ 0 & s_y \end{pmatrix} = \begin{pmatrix} s_x & s_y \sin ø \\ 0 & s_y \cos ø \end{pmatrix}$$

$$\begin{pmatrix} s_x & 0 \\ 0 & s_y \end{pmatrix} \begin{pmatrix} 1 & \sin ø \\ 0 & \cos ø \end{pmatrix} = \begin{pmatrix} s_x & s_x \sin ø \\ 0 & s_y \cos ø \end{pmatrix}$$

which is different. In the first case, we scale before we transform the coordinates, while in the second we apply the scale factor after the transformation.

MATRICES, DETERMINANTS, AND VECTORS

This sequencing is especially important when considering the rotation of axes in three dimensions. The measurement of angles in a clockwise direction is sometimes referred to as the left-hand rule because if your left thumb is pointing upwards (in the Z-direction) and the fingers move from straight to bent, their pointing moves clockwise as you look down on your thumb. If the index finger remains straight it represents the Y-direction.

In Figure 7.5(a), we have the axes X (east) Y (north) in the plane of the paper and Z perpendicular to the surface. Figures 7.5(b)–(d) show the effect of a clockwise turn about the X-axis, then Y, then Z, looking vertically down.

It should, however, be stressed that especially in photogrammetry, the opposite (right-hand) rule is used and positive rotations have the opposite sense. The confusion is compounded by whether an object is rotated about the axes or the axes are rotated about the object. For the present, we shall consider the rotation of axes; the key point is, as we have already observed, the sign of the sine.

If we are dealing with three dimensions (x, y, z), then the rotation matrix R about the Z-axis as shown in the two-dimensional case is

$$\mathbf{R}_z = \begin{pmatrix} \cos\phi_z & -\sin\phi_z & 0 \\ \sin\phi_z & \cos\phi_z & 0 \\ 0 & 0 & 1 \end{pmatrix}$$

where ϕ_z is the angle of clockwise rotation looking down the Z-axis. Similarly,

$$\mathbf{R}_x = \begin{pmatrix} 1 & 0 & 0 \\ 0 & \cos\phi_x & -\sin\phi_x \\ 0 & \sin\phi_x & \cos\phi_x \end{pmatrix} \quad \text{and} \quad \mathbf{R}_y = \begin{pmatrix} \cos\phi_y & 0 & -\sin\phi_y \\ 0 & 1 & 0 \\ \sin\phi_y & 0 & \cos\phi_y \end{pmatrix}$$

Thus, if the axes are rotated in the positive direction as defined in Figures 7.5(b), (c) and (d), then

$$\mathbf{X}_n = \begin{pmatrix} x_n \\ y_n \\ z_n \end{pmatrix} = \mathbf{R}_x \mathbf{R}_y \mathbf{R}_z \mathbf{X}_o$$

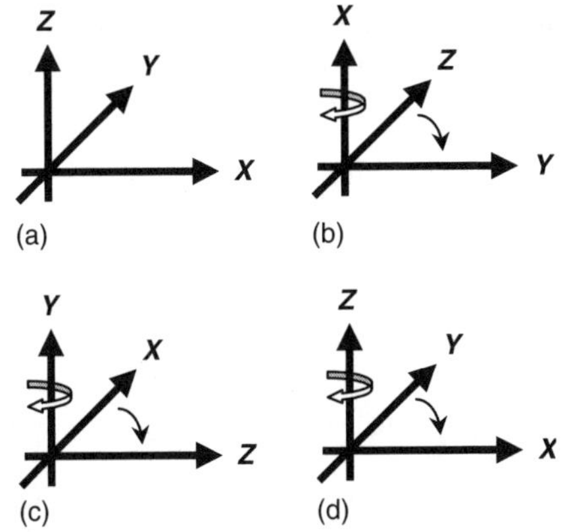

Figure 7.5 Positive rotations — left hand rule.

where $\mathbf{X}_o$ = the original coordinates

$$\begin{pmatrix} x_o \\ y_o \\ z_o \end{pmatrix}$$

It is important to keep in mind that the sequence in which the rotations take place is significant since in matrix algebra $\mathbf{A}*\mathbf{B}$ does not normally equal $\mathbf{B}*\mathbf{A}$. The rotations $\mathbf{R}_y \mathbf{R}_x \mathbf{R}_z$ in that order (applying $\mathbf{R}_x$ to $\mathbf{R}_z$ and then $\mathbf{R}_y$ to the result; or $\mathbf{R}_y$ to $\mathbf{R}_x$ and then the resulting product of $\mathbf{R}_y\mathbf{R}_x$ to $\mathbf{R}_z$) will give a different answer to that of $\mathbf{R}_x \mathbf{R}_y \mathbf{R}_z$. In the latter case, we rotate first about the z-axis, then about the y-axis, and then about the x (working from right to left).

Fortunately, $\mathbf{A}*\mathbf{B}*\mathbf{C} = \mathbf{A}*(\mathbf{B}*\mathbf{C}) = (\mathbf{A}*\mathbf{B})*\mathbf{C}$; thus, it does not matter whether we calculate $\mathbf{A}*\mathbf{B}$ first and then apply the result to $\mathbf{C}$ or whether we first calculate $\mathbf{B}*\mathbf{C}$ and then premultiply by $\mathbf{A}$.

In photogrammetry, the axis of the camera is said to be the z-axis and a rotation about it in the direction shown in Figure 7.6 is said to be "kappa" — the Greek letter κ. The rotations in Figure 7.6 are right handed in that if your right thumb points along the direction of the arrows your fingers curl to point as shown.

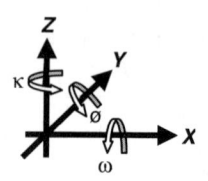

Figure 7.6 Photogrammetric rotations.

The direction of flight for a series of aerial photographs is called the x-axis and rotations of the camera about the x-axis are designated "omega" — the Greek letter ω. Rotations about the y-axis that is at right angles to the line of flight are said to be "phi" — the Greek letter ø. If we have a camera that we rotate first about the x-axis (known as the primary axis), then about the y-axis, and finally the z-axis, we will produce a transformation of the form

$$\mathbf{M} = \begin{pmatrix} \cos\kappa & \sin\kappa & 0 \\ -\sin\kappa & \cos\kappa & 0 \\ 0 & 0 & 1 \end{pmatrix} * \begin{pmatrix} \cos\phi & 0 & -\sin\phi \\ 0 & 1 & 0 \\ \sin\phi & 0 & \cos\phi \end{pmatrix} * \begin{pmatrix} 1 & 0 & 0 \\ 0 & \cos\omega & \sin\omega \\ 0 & -\sin\omega & \cos\omega \end{pmatrix}$$

$$= \begin{pmatrix} \cos\phi\cos\kappa & \cos\omega\sin\kappa + \sin\omega\sin\phi\cos\kappa & \sin\omega\sin\kappa - \cos\omega\sin\phi\cos\kappa \\ -\cos\phi\sin\kappa & \cos\omega\cos\kappa - \sin\omega\sin\phi\sin\kappa & \sin\omega\cos\kappa + \cos\omega\sin\phi\sin\kappa \\ \sin\phi & -\sin\omega\cos\phi & \cos\omega\cos\phi \end{pmatrix}$$

In particular, if the angles through which we make the rotations are small and we call these δω, δø, and δκ, then we have shown in Chapter 6 that cos δω, cos δø, and cos δκ each equal 1, while the sine functions become δω, δø, and δκ in radians.

MATRICES, DETERMINANTS, AND VECTORS

Thus, for small rotations, the matrix **M** reduces to

$$\begin{pmatrix} 1 & \delta\kappa & -\delta\phi \\ -\delta\kappa & 1 & \delta\omega \\ \delta\phi & -\delta\omega & 1 \end{pmatrix}$$

Further details can be found in standard books on photogrammetry. Here, our interest is in the mathematics behind the photogrammetric measurements.

7.6 SIMPLIFYING MATRICES

Matrices can be *partitioned*, that is, treated in parts. Consider the two matrices **A** with 4 rows and 5 columns and **B** with 5 rows and 2 columns.

$$\mathbf{A} * \mathbf{B} = \begin{pmatrix} a_{11} & a_{12} & a_{13} & a_{14} & a_{15} \\ a_{21} & a_{22} & a_{23} & a_{24} & a_{25} \\ a_{31} & a_{32} & a_{33} & a_{34} & a_{35} \\ a_{41} & a_{42} & a_{43} & a_{44} & a_{45} \end{pmatrix} * \begin{pmatrix} b_{11} & b_{12} \\ b_{21} & b_{22} \\ b_{31} & b_{32} \\ b_{41} & b_{42} \\ b_{51} & b_{52} \end{pmatrix}$$

which we can write in the form

$$\mathbf{A} * \mathbf{B} = \left(\begin{array}{ccc|cc} a_{11} & a_{12} & a_{13} & a_{14} & a_{15} \\ a_{21} & a_{22} & a_{23} & a_{24} & a_{25} \\ a_{31} & a_{32} & a_{33} & a_{34} & a_{35} \\ a_{41} & a_{42} & a_{43} & a_{44} & a_{45} \end{array}\right) * \left(\begin{array}{cc} b_{11} & b_{12} \\ b_{21} & b_{22} \\ b_{31} & b_{32} \\ \hline b_{41} & b_{42} \\ b_{51} & b_{52} \end{array}\right)$$

or

$$\mathbf{A} * \mathbf{B} = (\mathbf{A}_1 \quad \mathbf{A}_2) * \begin{pmatrix} \mathbf{B}_1 \\ \mathbf{B}_2 \end{pmatrix}$$

where

$$\mathbf{A}_1 = \begin{pmatrix} a_{11} & a_{12} & a_{13} \\ a_{21} & a_{22} & a_{23} \\ a_{31} & a_{32} & a_{33} \\ a_{41} & a_{42} & a_{43} \end{pmatrix}, \mathbf{A}_2 = \begin{pmatrix} a_{14} & a_{15} \\ a_{24} & a_{25} \\ a_{34} & a_{35} \\ a_{44} & a_{45} \end{pmatrix}, \mathbf{B}_1 = \begin{pmatrix} b_{11} & b_{12} \\ b_{21} & b_{22} \\ b_{31} & b_{32} \end{pmatrix}$$

and

$$\mathbf{B}_2 = \begin{pmatrix} b_{41} & b_{42} \\ b_{51} & b_{52} \end{pmatrix}$$

$\mathbf{A}_1$, $\mathbf{A}_2$, $\mathbf{B}_1$, and $\mathbf{B}_2$ are partitions of the matrix.
The original multiplication $\mathbf{A} * \mathbf{B}$ is the same as

$$\begin{pmatrix} \mathbf{A}_1 \mathbf{B}_1 & \mathbf{A}_1 \mathbf{B}_2 \\ \mathbf{A}_2 \mathbf{B}_1 & \mathbf{A}_2 \mathbf{B}_2 \end{pmatrix}$$

We can also simplify matrices by making them triangular, where $\mathbf{A}_U$ is called an *upper triangular matrix*, while $\mathbf{A}_L$ is a *lower triangular matrix*:

$$\mathbf{A}_U = \begin{pmatrix} a_{11} & a_{12} & a_{13} & a_{14} & a_{15} \\ 0 & a_{22} & a_{23} & a_{24} & a_{25} \\ 0 & 0 & a_{33} & a_{34} & a_{35} \\ 0 & 0 & 0 & a_{44} & a_{45} \\ 0 & 0 & 0 & 0 & a_{55} \end{pmatrix}, \quad \mathbf{A}_L = \begin{pmatrix} a_{11} & 0 & 0 & 0 & 0 \\ a_{21} & a_{22} & 0 & 0 & 0 \\ a_{31} & a_{32} & a_{33} & 0 & 0 \\ a_{41} & a_{42} & a_{43} & a_{44} & 0 \\ a_{51} & a_{52} & a_{53} & a_{54} & a_{55} \end{pmatrix}$$

To illustrate how this works, consider five equations for five unknowns $(x_1, x_2, x_3, x_4, x_5)$

$$a_{11}x_1 + a_{12}x_2 + a_{13}x_3 + a_{14}x_4 + a_{15}x_5 = b_1$$

$$a_{21}x_2 + a_{22}x_2 + a_{23}x_3 + a_{24}x_4 + a_{25}x_5 = b_2$$

$$a_{31}x_3 + a_{32}x_2 + a_{33}x_3 + a_{34}x_4 + a_{35}x_5 = b_3$$

$$a_{41}x_4 + a_{42}x_2 + a_{43}x_3 + a_{44}x_4 + a_{45}x_5 = b_4$$

$$a_{51}x_5 + a_{52}x_2 + a_{53}x_3 + a_{54}x_4 + a_{55}x_5 = b_5$$

where a_{11}, a_{12}, etc., and b_1, b_2, etc., are known quantities.

We can express this in the form

$$\begin{pmatrix} a_{11} & a_{12} & a_{13} & a_{14} & a_{15} \\ a_{21} & a_{22} & a_{23} & a_{24} & a_{25} \\ a_{31} & a_{32} & a_{33} & a_{34} & a_{35} \\ a_{41} & a_{42} & a_{43} & a_{44} & a_{45} \\ a_{51} & a_{52} & a_{53} & a_{54} & a_{55} \end{pmatrix} * \begin{pmatrix} x_1 \\ x_2 \\ x_3 \\ x_4 \\ x_5 \end{pmatrix} = \begin{pmatrix} b_1 \\ b_2 \\ b_3 \\ b_4 \\ b_5 \end{pmatrix}$$

or as $\mathbf{AX} = \mathbf{B}$.

Now premultiply both sides of $\mathbf{AX} = \mathbf{B}$ by the diagonal matrix $\mathbf{C}_1$, whose leading diagonal is made from the reciprocals of the numbers in column 1, that is, with diagonal with elements $\{1/a_{11}, 1/a_{21}, 1/a_{31}, 1/a_{41}, 1/a_{51}\}$.

We obtain $\mathbf{C}_1\mathbf{AX} = \mathbf{C}_1\mathbf{B}$ or

$$\begin{pmatrix} a_{11}/a_{11} & a_{12}/a_{11} & a_{13}/a_{11} & a_{14}/a_{11} & a_{15}/a_{11} \\ a_{21}/a_{21} & a_{22}/a_{21} & a_{23}/a_{21} & a_{24}/a_{21} & a_{25}/a_{21} \\ a_{31}/a_{31} & a_{32}/a_{31} & a_{33}/a_{31} & a_{34}/a_{31} & a_{35}/a_{31} \\ a_{41}/a_{41} & a_{42}/a_{41} & a_{43}/a_{41} & a_{44}/a_{41} & a_{45}/a_{41} \\ a_{51}/a_{51} & a_{52}/a_{51} & a_{53}/a_{51} & a_{54}/a_{51} & a_{55}/a_{51} \end{pmatrix} * \begin{pmatrix} x_1 \\ x_2 \\ x_3 \\ x_4 \\ x_5 \end{pmatrix} = \begin{pmatrix} b_1/a_{11} \\ b_2/a_{21} \\ b_3/a_{31} \\ b_4/a_{41} \\ b_5/a_{51} \end{pmatrix}$$

or

$$\begin{pmatrix} 1 & a_{12}/a_{11} & a_{13}/a_{11} & a_{14}/a_{11} & a_{15}/a_{11} \\ 1 & a_{22}/a_{21} & a_{23}/a_{21} & a_{24}/a_{21} & a_{25}/a_{21} \\ 1 & a_{32}/a_{31} & a_{33}/a_{31} & a_{34}/a_{31} & a_{35}/a_{31} \\ 1 & a_{42}/a_{41} & a_{43}/a_{41} & a_{44}/a_{41} & a_{45}/a_{41} \\ 1 & a_{52}/a_{51} & a_{53}/a_{51} & a_{54}/a_{51} & a_{55}/a_{51} \end{pmatrix} * \begin{pmatrix} x_1 \\ x_2 \\ x_3 \\ x_4 \\ x_5 \end{pmatrix} = \begin{pmatrix} b_1/a_{11} \\ b_2/a_{21} \\ b_3/a_{31} \\ b_4/a_{41} \\ b_5/a_{51} \end{pmatrix}$$

MATRICES, DETERMINANTS, AND VECTORS   107

Subtract row 1 from the remaining rows to give

$$\begin{pmatrix} 1 & a_{12}/a_{11} & a_{13}/a_{11} & a_{14}/a_{11} & a_{15}/a_{11} \\ 0 & a_{22}/a_{21} - a_{12}/a_{11} & a_{23}/a_{21} - a_{13}/a_{11} & a_{24}/a_{21} - a_{14}/a_{11} & a_{25}/a_{21} - a_{15}/a_{11} \\ 0 & a_{32}/a_{31} - a_{12}/a_{11} & a_{33}/a_{31} - a_{13}/a_{11} & a_{34}/a_{31} - a_{14}/a_{11} & a_{35}/a_{31} - a_{15}/a_{11} \\ 0 & a_{42}/a_{41} - a_{12}/a_{11} & a_{43}/a_{41} - a_{13}/a_{11} & a_{44}/a_{41} - a_{14}/a_{11} & a_{45}/a_{41} - a_{15}/a_{11} \\ 0 & a_{52}/a_{51} - a_{12}/a_{11} & a_{53}/a_{51} - a_{13}/a_{11} & a_{54}/a_{51} - a_{14}/a_{11} & a_{55}/a_{51} - a_{15}/a_{11} \end{pmatrix}$$

$$* \begin{pmatrix} x_1 \\ x_2 \\ x_3 \\ x_4 \\ x_5 \end{pmatrix} = \begin{pmatrix} b_1/a_{11} \\ b_2/a_{21} - b_1/a_{11} \\ b_3/a_{31} - b_1/a_{11} \\ b_4/a_{41} - b_1/a_{11} \\ b_5/a_{51} - b_1/a_{11} \end{pmatrix}$$

We can then repeat this process ignoring the first row and first column and considering only the (4×4) matrix of rows 2 to 5 and columns 2 to 5 (i.e., we consider the new submatrix of the original row one column one). We then repeat this again on the (3×3) and again on the (2×2), finishing up with an upper triangular matrix

$$\begin{pmatrix} 1 & a_{12}/a_{11} & a_{13}/a_{11} & a_{14}/a_{11} & a_{15}/a_{11} \\ 0 & 1 & a'_{23} & a'_{24} & a'_{25} \\ 0 & 0 & 1 & a'_{34} & a'_{35} \\ 0 & 0 & 0 & 1 & a'_{45} \\ 0 & 0 & 0 & 0 & 1 \end{pmatrix} * \begin{pmatrix} x_1 \\ x_2 \\ x_3 \\ x_4 \\ x_5 \end{pmatrix} = \begin{pmatrix} b_1/a_{11} \\ b'_2 \\ b'_3 \\ b'_4 \\ b'_5 \end{pmatrix}$$

The process is boring and repetitive, ideally suited to automate in a computer program that can routinely evaluate all the a' and b' (Box 7.8).

Box 7.8 — Example of Solutions of Simultaneous Equations

Let

$$x + y + z + w = 11$$

$$2x - 6y + 3z + 7w = 24$$

$$7x + 2y + 5z - 3w = 10$$

$$9x - 3y + 2z + 2w = 21$$

In matrix form

$$\begin{pmatrix} 1 & 1 & 1 & 1 \\ 2 & -6 & 3 & 7 \\ 7 & 2 & 5 & -3 \\ 9 & -3 & 2 & 2 \end{pmatrix} * \begin{pmatrix} x \\ y \\ z \\ w \end{pmatrix} = \begin{pmatrix} 11 \\ 24 \\ 10 \\ 21 \end{pmatrix} \quad \text{or}$$

—Continued

Box 7.8 — Example of Solutions of Simultaneous Equations (Continued)

$$\begin{pmatrix} 1 & 1 & 1 & 1 \\ 1 & -3 & 1.5 & 3.5 \\ 1 & 0.286 & 0.714 & -0.429 \\ 1 & -0.333 & 0.222 & 0.222 \end{pmatrix} * \begin{pmatrix} x \\ y \\ z \\ w \end{pmatrix} = \begin{pmatrix} 11 \\ 12 \\ 1.429 \\ 2.333 \end{pmatrix}$$

or

$$\begin{pmatrix} 1 & 1 & 1 & 1 \\ 0 & -4 & 0.5 & 2.5 \\ 0 & -0.714 & -0.286 & -1.429 \\ 0 & -1.333 & -0.778 & -0.778 \end{pmatrix} * \begin{pmatrix} x \\ y \\ z \\ w \end{pmatrix} = \begin{pmatrix} 11 \\ 1 \\ -9.571 \\ -8.667 \end{pmatrix}$$

or

$$\begin{pmatrix} 1 & 1 & 1 & 1 \\ 0 & 1 & -0.125 & -0.625 \\ 0 & 1 & 0.4 & 2 \\ 0 & 1 & 0.584 & 0.584 \end{pmatrix} * \begin{pmatrix} x \\ y \\ z \\ w \end{pmatrix} = \begin{pmatrix} 11 \\ -0.25 \\ 13.4 \\ 6.5 \end{pmatrix}$$

or

$$\begin{pmatrix} 1 & 1 & 1 & 1 \\ 0 & 1 & -0.125 & -0.625 \\ 0 & 0 & 0.525 & 2.625 \\ 0 & 0 & 0.71 & 1.21 \end{pmatrix} * \begin{pmatrix} x \\ y \\ z \\ w \end{pmatrix} = \begin{pmatrix} 11 \\ -0.25 \\ 13.65 \\ 6.75 \end{pmatrix}$$

or

$$\begin{pmatrix} 1 & 1 & 1 & 1 \\ 0 & 1 & -0.125 & -0.625 \\ 0 & 0 & 1 & 5 \\ 0 & 0 & 1 & 1.7 \end{pmatrix} * \begin{pmatrix} x \\ y \\ z \\ w \end{pmatrix} = \begin{pmatrix} 11 \\ -0.25 \\ 26 \\ 9.5 \end{pmatrix}$$

Hence

$$\begin{pmatrix} 1 & 1 & 1 & 1 \\ 0 & 1 & -0.125 & -0.625 \\ 0 & 0 & 1 & 5 \\ 0 & 0 & 0 & -3.3 \end{pmatrix} * \begin{pmatrix} x \\ y \\ z \\ w \end{pmatrix} = \begin{pmatrix} 11 \\ -0.25 \\ 26 \\ -16.5 \end{pmatrix}$$

or

$$\begin{pmatrix} 1 & 1 & 1 & 1 \\ 0 & 1 & -0.125 & -0.625 \\ 0 & 0 & 1 & 5 \\ 0 & 0 & 0 & 1 \end{pmatrix} * \begin{pmatrix} x \\ y \\ z \\ w \end{pmatrix} = \begin{pmatrix} 11 \\ -0.25 \\ 26 \\ 5 \end{pmatrix}$$

Thus, $w = 5$

$$z + 5w = 26 \quad \text{so} \quad z = 1$$

$$y - 0.125z - 0.625w = -0.25 \quad \text{so} \quad y = 3$$

$$x + y + z + w = 11 \quad \text{so} \quad x = 2$$

MATRICES, DETERMINANTS, AND VECTORS   109

The numbers **X** are found by back substitution. Thus, $x_5 = b'_5$, the number that one finishes up with in the last row of the modified matrix **B′**. Moving up a row in matrix **B′**, $x_4 + a'_{45}x_5 = b'_4$ or

$x_4 = b'_4 - a'_{45}x_5$ and so on

If we return for a moment to the original expression of **AX** = **B**, namely

$$\begin{pmatrix} a_{11} & a_{12} & a_{13} & a_{14} & a_{15} \\ a_{21} & a_{22} & a_{23} & a_{24} & a_{25} \\ a_{31} & a_{32} & a_{33} & a_{34} & a_{35} \\ a_{41} & a_{42} & a_{43} & a_{44} & a_{45} \\ a_{51} & a_{52} & a_{53} & a_{54} & a_{55} \end{pmatrix} * \begin{pmatrix} x_1 \\ x_2 \\ x_3 \\ x_4 \\ x_5 \end{pmatrix} = \begin{pmatrix} b_1 \\ b_2 \\ b_3 \\ b_4 \\ b_5 \end{pmatrix}$$

The manipulation was all in terms of '*a*' and '*b*'. We could have written the equation in one solid block of numbers as

$$\begin{pmatrix} a_{11} & a_{12} & a_{13} & a_{14} & a_{15} & b_1 \\ a_{21} & a_{22} & a_{23} & a_{24} & a_{25} & b_2 \\ a_{31} & a_{32} & a_{33} & a_{34} & a_{35} & b_3 \\ a_{41} & a_{42} & a_{43} & a_{44} & a_{45} & b_4 \\ a_{51} & a_{52} & a_{53} & a_{54} & a_{55} & b_5 \end{pmatrix} \text{ in which } \begin{pmatrix} b_1 \\ b_2 \\ b_3 \\ b_4 \\ b_5 \end{pmatrix}$$

is a column *vector of constants*, while the '*a*' terms form the *matrix of coefficients*. The combined expression is known as an *augmented matrix*.

When reducing a matrix to diagonal form by the method outlined above, a problem can arise where one of the coefficients used for division is zero. This will happen when the five equations are not independent in which case there is no solution — there must at least be the same number of independent equations as there are unknowns in order to find a unique solution. If there are extra equations as can happen with survey measurements, where there are redundant measurements in order to improve the accuracy of the result, then a most probable solution can be found using statistical techniques such as the least square solution that is discussed in Chapter 11.

7.7 VECTORS

Although closely related, matrix algebra should not be confused with the handling of vectors. A *vector* is a quantity such as velocity that has both magnitude and direction, while a *scalar* is a quantity that has magnitude but no direction.

The magnitude may represent a true distance such as a length or a quantity such as force or speed. In the study of geographic information systems, there is a tendency to think of vectors as the lengths and directions of lines and compare them with raster representation. In practice, they may represent the strength and direction of a variety of phenomena such as gravity. A bearing combined with a distance is only one example of a vector.

OP in Figure 7.7 represents a vector based on orthogonal axes that could be called *x*, *y*, and *z* but which we will call the directions **i**, **j**, **k**. They are written in bold

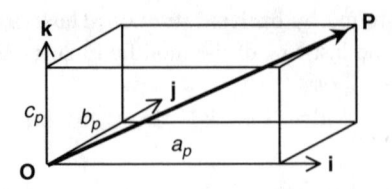

Figure 7.7 The axes **i**, **j** and **k** for vector **P**.

letters (rather like writing the symbol for the matrix **M** in bold script) to identify them as *unit vectors*, that is, vectors of unit length. If the actual length of the sides of a rectangular block of which **OP** is a diagonal are a_p, b_p, and c_p, then we can express the vector **OP** as follows:

$$\mathbf{OP} = a_p \mathbf{i} + b_p \mathbf{j} + c_p \mathbf{k}.$$

If we have another vector **PQ** starting at **P** as in Figure 7.8 such that

$$\mathbf{PQ} = a_q \mathbf{i} + b_q \mathbf{j} + c_q \mathbf{k}$$

then

$$\mathbf{OQ} = (a_p + a_q)\mathbf{i} + (b_p + b_q)\mathbf{j} + (c_p + c_q)\mathbf{k}$$

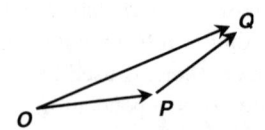

Figure 7.8 Vector addition.

From Q to O, one needs to travel negatively in each of the directions **i**, **j**, **k**. In fact,

$$\mathbf{OP} + \mathbf{PQ} + \mathbf{QO} = 0$$

We can also describe a vector in two parts — its length and its direction in the form $\mathbf{a} = |\mathbf{a}|\,\hat{\mathbf{a}}$.

This means that the vector **a** has length determined by its modulus $|\mathbf{a}|$ and the direction expressed as $\hat{\mathbf{a}}$. When $\mathbf{OP} = a_p\mathbf{i} + b_p\mathbf{j} + c_p\mathbf{k}$, the length of OP can be calculated by Pythagoras to be $|\mathbf{OP}| = \sqrt{(a_p^2 + b_p^2 + c_p^2)}$. Note that $|\mathbf{OP}|$ (the modulus of vector **OP**) is a scalar quantity and has no direction.

The cosines of the angles that **OP** makes with the reference vectors **i**, **j**, and **k** are called the *direction cosines* and are marked as α (angle **POi**), β (angle **POj**) and γ (angle **POk**) in Figure 7.9:

$$a_p = |\mathbf{OP}|\cos\alpha, \quad b_p = |\mathbf{OP}|\cos\beta \quad c_p = |\mathbf{OP}|\cos\gamma.$$

Thus, a vector can be defined by its length and the three direction cosines.

MATRICES, DETERMINANTS, AND VECTORS

The length of a vector can be changed simply by applying a scale factor so that, for example,

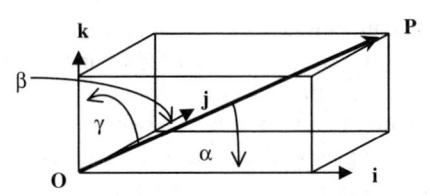

$$3\mathbf{a} = 3\,|\mathbf{a}|\,\hat{\mathbf{a}} \quad \text{so if} \quad \mathbf{a} = 2\mathbf{i} + 3\mathbf{j} + 4\mathbf{k} \quad \text{then} \quad 3\mathbf{a} = 6\mathbf{i} + 9\mathbf{j} + 12\mathbf{k}$$

Figure 7.9 Direction cosines.

Vectors can be added and subtracted. They can also be multiplied or divided by a scalar quantity. Two vectors can also be multiplied; but the concept has a meaning that is different from that of ordinary multiplication. Consider two such vectors **a** and **b**, such that

$$\mathbf{a} = a_x\mathbf{i} + a_y\mathbf{j} + a_z\mathbf{k} \quad \text{and} \quad \mathbf{b} = b_x\mathbf{i} + b_y\mathbf{j} + b_z\mathbf{k}$$

There are two forms of product, known as the *dot product* and the *cross product*.

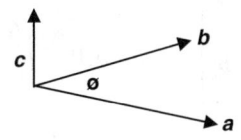

Figure 7.10 Dot and cross products.

The dot product results in a scalar quantity defined as

$$\mathbf{a} \cdot \mathbf{b} = |\mathbf{a}||\mathbf{b}|\cos \emptyset$$

where ø is the difference in direction or angle between the two vectors. In particular, since $\cos(0) = 1$ and $\cos(90°) = 0$

$$\mathbf{i} \cdot \mathbf{i} = \mathbf{j} \cdot \mathbf{j} = \mathbf{k} \cdot \mathbf{k} = 1 \quad \text{while} \quad \mathbf{i} \cdot \mathbf{j} = \mathbf{j} \cdot \mathbf{k} = \mathbf{k} \cdot \mathbf{i} = 0$$

Thus, $\mathbf{a} \cdot \mathbf{b} = (a_x\mathbf{i} + a_y\mathbf{j} + a_z\mathbf{k}) \cdot (b_x\mathbf{i} + b_y\mathbf{j} + b_z\mathbf{k})$ giving $3*3 = 9$ relationships that contain terms such as $\mathbf{i} \cdot \mathbf{i}$ (which $= 1$) or $\mathbf{i} \cdot \mathbf{j}$ (which $= 0$) or $\mathbf{j} \cdot \mathbf{i}$ (which also $= 0$). Working this through (Box 7.9),

$$\mathbf{a} \cdot \mathbf{b} = a_x b_x + a_y b_y + a_z b_z$$

The cross product creates a new vector perpendicular to the plane containing **a** and **b** (shown by the vector **c** in Figure 7.10). Its origin lies in the study of electromagnetism, although it has relevance to a number of problems in geomatics, for instance, when determining whether any surface element of a feature, such as a building façade, is facing toward or away from the observer. This allows levels of shading to be calculated or hidden surfaces to be removed. The cross product of **a** × **b** = |**a**||**b**| sin ø**c**. The magnitude = |**a**||**b**| sin ø, which is a scalar quantity and the direction is determined by convention to be a right-handed system.

It follows that if we reverse the order of multiplication, the vector **c** will be pointing in the opposite direction that is, **a** × **b** = −**b** × **a**.

For unit vectors, the angles between them are 90° and sin (90°) = 1 and cos (90°) = 0. Hence, **i** × **i** = **j** × **j** = **k** × **k** = 0. Also,

$$\mathbf{i} \times \mathbf{j} = \mathbf{k}; \mathbf{j} \times \mathbf{k} = \mathbf{i}; \mathbf{k} \times \mathbf{i} = \mathbf{j} \text{ and } \mathbf{j} \times \mathbf{i} = -\mathbf{k}; \mathbf{k} \times \mathbf{j} = -\mathbf{i} \text{ and } \mathbf{i} \times \mathbf{k} = -\mathbf{j}$$

If as before we put $\mathbf{a} = a_x\mathbf{i} + a_y\mathbf{j} + a_z\mathbf{k}$ and $\mathbf{b} = b_x\mathbf{i} + b_y\mathbf{j} + b_z\mathbf{k}$, then $\mathbf{a} \times \mathbf{b} = (a_x\mathbf{i} + a_y\mathbf{j} + a_z\mathbf{k}) \times (b_x\mathbf{i} + b_y\mathbf{j} + b_z\mathbf{k})$ giving 3∗3 = 9 relationships that contain terms such as **i** × **i** (which = 0) or **i** × **j** (which = **k**) or **j** × **i** (which = −**k**). Working this through,

$$\mathbf{a} \times \mathbf{b} = (a_y b_z - a_z b_y)\mathbf{i} + (a_z b_x - a_x b_z)\mathbf{j} = (a_x b_y - a_y b_x)\mathbf{k}$$

Or to express this in determinant form

$$\mathbf{a} \times \mathbf{b} = \begin{vmatrix} \mathbf{i} & \mathbf{j} & \mathbf{k} \\ a_x & a_y & a_z \\ b_x & b_y & b_z \end{vmatrix} = \begin{vmatrix} a_y & a_z \\ b_y & b_z \end{vmatrix} \mathbf{i} - \begin{vmatrix} a_x & a_z \\ b_x & b_z \end{vmatrix} \mathbf{j} + \begin{vmatrix} a_x & a_y \\ b_x & b_y \end{vmatrix} \mathbf{k}$$

The relation **a** . **b** × **c** represents (Box 7.9) the scalar product between two vectors, the vector **a** and the vector **b** × **c**. It is called the *scalar triple product* and gives the volume of the solid whose edges are represented by the vectors **a**, **b**, and **c**. This solid has parallel sides and is in effect a squashed brick and is known as a *parallelepiped* (Figure 7.11).

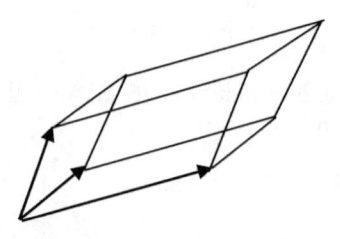

Figure 7.11 A parallelepiped.

If $\mathbf{a} = a_x\mathbf{i} + a_y\mathbf{j} + a_z\mathbf{k}$; $\mathbf{b} = b_x\mathbf{i} + b_y\mathbf{j} + b_z\mathbf{k}$; and $\mathbf{c} = c_x\mathbf{i} + c_y\mathbf{j} + c_z\mathbf{k}$, then $\mathbf{b} \times \mathbf{c} = (b_y c_z - b_z c_y)\mathbf{i} + (b_z c_x - b_x c_z)\mathbf{j} + (b_x c_y - b_y c_x)\mathbf{k}$. Hence,

MATRICES, DETERMINANTS, AND VECTORS

$\mathbf{a} \cdot (\mathbf{b} \times \mathbf{c})$
$= \{a_x\mathbf{i} + a_y\mathbf{j} + a_z\mathbf{k}\} \cdot \{(b_yc_z - b_zc_y)\mathbf{i} + (b_zc_x - b_xc_z)\mathbf{j} + (b_xc_y - b_yc_x)\mathbf{k}\}$
$= a_x(b_yc_z - b_zc_y) + a_y(b_zc_x - b_xc_z) + a_z(b_xc_y - b_yc_x)$
$= a_x(b_yc_z - b_zc_y) - a_y(b_xc_z - b_zc_x) + a_z(b_xc_y - b_yc_x)$

Thus

$$\mathbf{a} \cdot (\mathbf{b} \times \mathbf{c}) = \begin{vmatrix} a_x & a_y & a_z \\ b_x & b_y & b_z \\ c_x & c_y & c_z \end{vmatrix}$$

Box 7.9 — Vector Multiplication

If $\mathbf{a} = a_x\mathbf{i} + a_y\mathbf{j} + a_z\mathbf{k}$ and $\mathbf{b} = b_x\mathbf{i} + b_y\mathbf{j} + b_z\mathbf{k}$, then

$$\mathbf{a} \cdot \mathbf{b} = a_xb_x + a_yb_y + a_zb_z$$

$$\mathbf{a} \times \mathbf{b} = (a_yb_z - a_zb_y)\mathbf{i} + (a_zb_x - a_xb_z)\mathbf{j} + (a_xb_y - a_yb_x)\mathbf{k}$$

$$\mathbf{a} \cdot (\mathbf{b} \times \mathbf{c}) = \begin{vmatrix} a_x & a_y & a_z \\ b_x & b_y & b_z \\ c_x & c_y & c_z \end{vmatrix}$$

More importantly, if we have three points:

$$A\ (x_A, y_A, z_A),\ B\ (x_B, y_B, z_B),\ \text{and}\ C\ (x_C, y_C, z_C)$$

then these are three vectors (Figure 7.12) from the origin $P\ (0, 0, 0)$.

$$\mathbf{P}_A = (x_A\mathbf{i}, y_A\mathbf{j}, z_A\mathbf{k}),\ \mathbf{P}_B = (x_B\mathbf{i}, y_B\mathbf{j}, z_B\mathbf{k}),\ \text{and}\ \mathbf{P}_C = (x_C\mathbf{i}, y_C\mathbf{j}, z_C\mathbf{k})$$

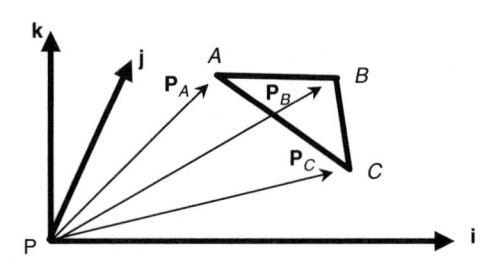

Figure 7.12 Vectors and a plane.

The lines CA and CB are represented by $(\mathbf{P}_A - \mathbf{P}_C)$ and $(\mathbf{P}_B - \mathbf{P}_C)$. The cross product between these two lines is given by.

$$(\mathbf{P}_A - \mathbf{P}_C) \times (\mathbf{P}_B - \mathbf{P}_C)$$
$$= \{(x_A - x_C)\mathbf{i} + (y_A - y_C)\mathbf{j} + (z_A - z_C)\mathbf{k}\}$$
$$\times \{(x_B - x_C)\mathbf{i} + (y_B - y_C)\mathbf{j} + (z_B - z_C)\mathbf{k}\}$$
$$= \{(x_A - x_C)(y_B - y_C)\mathbf{k} - (x_A - x_C)(z_B - z_C)\mathbf{j} - (y_A - y_C)(x_B - x_C)\mathbf{k}$$
$$+ (y_A - y_C)(z_B - z_C)\mathbf{i} + (z_A - z_C)(x_B - x_C)\mathbf{j} - (z_A - z_C)(y_B - y_C)\mathbf{i}\}$$
$$= \{y_A(z_B - z_C) + y_B(z_C - z_A) + y_C(z_A - z_B)\}\mathbf{i}$$
$$+ \{z_A(x_B - x_C) + z_B(x_C - x_A) + z_C(x_A - x_B)\}\mathbf{j}$$
$$+ \{x_A(y_B - y_C) + x_B(y_C - y_A) + x_C(y_A - y_B)\}\mathbf{k}$$

This represents a vector that defines the normal to the plane of CAB. If we join any general point $Q(x, y, z)$ to C, then the line CQ will be $(\mathbf{P}_Q - \mathbf{P}_C)$. The dot product with the above will be zero if the point Q lies in the plane ABC. That is, for Q to lie in the plane ABC

$$(\mathbf{P}_A - \mathbf{P}_C) \times (\mathbf{P}_B - \mathbf{P}_C) \cdot (\mathbf{P}_Q - \mathbf{P}_C) = 0$$

or

$$(\mathbf{P}_A - \mathbf{P}_C) \times (\mathbf{P}_B - \mathbf{P}_C) \cdot \{(x - x_C)\mathbf{i} + (y - y_C)\mathbf{j} + (z - z_C)\mathbf{k}\}$$

This gives

$$(x - x_C)\{y_A(z_B - z_C) + y_B(z_C - z_A) + y_C(z_A - z_B)\}$$
$$+ (y - y_C)\{z_A(x_B - x_C) + z_B(x_C - x_A) + z_C(x_A - x_B)\}$$
$$+ (z - z_C)\{x_A(y_B - y_C) + x_B(y_C - y_A) + x_C(y_A - y_B)\}$$
$$= 0$$

This is the equation of the plane ABC (Box 7.10).

Box 7.10 — The Equation for a Plane

The plane that contains three points:

$$A(x_A, y_A, z_A), B(x_B, y_B, z_B), \text{ and } C(x_C, y_C, z_C)$$

is given by

$$(x - x_C)\{y_A(z_B - z_C) + y_B(z_C - z_A) + y_C(z_A - z_B)\}$$
$$+ (y - y_C)\{z_A(x_B - x_C) + z_B(x_C - x_A) + z_C(x_A - x_B)\}$$
$$+ (z - z_C)\{x_A(y_B - y_C) + x_B(y_C - y_A) + x_C(y_A - y_B)\} = 0$$

CHAPTER 8

Curves and Surfaces

CONTENTS

8.1 Parametric Forms 115
8.2 The Ellipse 119
8.3 The Radius of Curvature 121
8.4 Fitting Curves to Points 122
8.5 The Bezier Curve 128

8.1 PARAMETRIC FORMS

So far, we have seen that for points on a plane, we can express a straight line as a series of values (x, y), such that

$$ax + by + c = 0$$

We can also express it in the following form:

$$y = mx + n$$

where 'm' is the slope or *gradient* of the line and 'n' is a constant. These two expressions represent the same line if

$$m = -a/b \text{ and } n = -c/b$$

We also saw in Chapter 6 and can see from Figure 8.1 that if the slope of the line AB is θ (measured from the horizontal axis anticlockwise), then $y = x \tan \theta + n$.

We can also express this as $dy/dx = \tan \theta = m$, which is constant for a straight line. So $d^2y/dx^2 = 0$. Also note that if we use clockwise bearings from the vertical rather than anticlockwise angles from the horizontal, the bearing of line $AB = (90 - \theta)$ and $y = x \cot (\text{Bearing } AB) + n$.

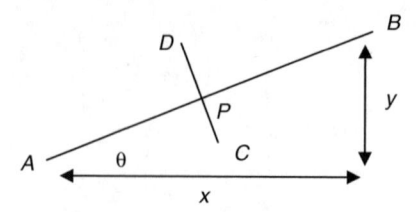

Figure 8.1 Orthogonal lines.

If the line *CD* is perpendicular to *AB*, its slope will be $(90 + \theta)$. Hence, for the line $CD: y = m'x + n'$ where

$$m' = -\cot(\theta) = -\tan(\text{Bearing } AB)$$

We have also seen that when two lines *AB* and *CD* are at right angles, their slopes are such that $m\,m' = \tan\theta * (-\cot\theta) = -1$. If the product of the slopes of two lines $= -1$, they are said to be *orthogonal*. In Chapter 7, we also used the term "orthogonal" for a matrix where $\mathbf{A}\,\mathbf{A}^{-1} = \mathbf{I} = \mathbf{A}\,\mathbf{A}^T$.

In the case of curves, we have seen that the slope at any point on the curve is dy/dx and the normal to it will have a slope $(-dx/dy)$. The straight line that represents the slope at any point is called a tangent. In the case of a point on a circle, the tangent is normal to the radius at that point. In the case of the tangent to an ellipse, the normal does not pass through the center except when the tangent is parallel to one of the axes (Figure 8.2).

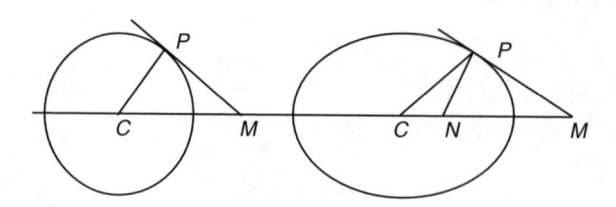

Figure 8.2 Tangents to a circle and an ellipse.

As we have already seen, the equation for a circle with center at the point (x_c, y_c) can be expressed in the form $(x - x_c)^2 + (y - y_c)^2 = r^2$.

An alternative way would be

$$x = r\cos\theta + x_c$$

$$y = r\sin\theta + y_c$$

where the variable θ (measured from the horizontal) is such that

$$0 \leq \theta \leq 2\pi.$$

Consider the ellipse in Figure 8.3 with center *O*, semi major axis '*a*,' and semi-minor axis '*b*.' As we saw in Chapter 4, the circle of radius '*a*' centered on *O* is called

CURVES AND SURFACES

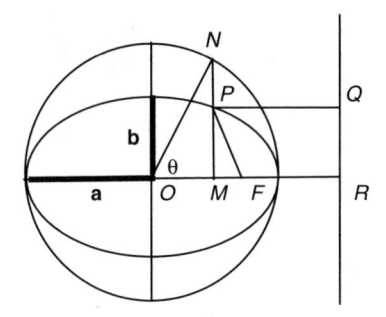

Figure 8.3 The ellipse.

the *auxiliary circle*. In effect, the auxiliary circle is squashed or scaled down in the vertical or y-direction to give the ellipse. Thus, if N is a point on the auxiliary circle, then we can regard it as having coordinates

$$x_n = a \cos \theta + x_c, \quad y_n = a \sin \theta + y_c$$

The length $MN = a \sin \theta$. By squashing the auxiliary circle such that OM remains the same but MP becomes scaled down so that $MP/MN = b/a$

$$MP = (b/a){*}a \sin \theta = b \sin \theta$$

Hence, for P on the ellipse,

$$x_p = a \cos \theta + x_c$$

$$y_p = b \sin \theta + y_c$$

the center of the ellipse O being (x_c, y_c).

Then, $\quad \sin^2 \theta + \cos^2 \theta = 1 = (x_p - x_c)^2/a^2 + (y_p - y_c)^2/b^2.$

To pursue the ellipse a little further, the distance to the directrix RQ was given in Chapter 4 as (a/e), with the value of 'e' being $0 < e < 1$. As for any conic, the distance $OF = a{*}e$, $OR = a/e$ and $FP = e{*}PQ$. For convenience, we will assume that the center of the ellipse is taken as the origin.
Since the lengths $PQ = MR$ and $MR + OM = a/e$, $PQ = a/e - x$.
Hence, $FP = e{*}PQ = a - ex$. $FM = OF - OM = ae - x$, while $MP = y$.
Also, $FP^2 = FM^2 + MP^2$. So

$$(a - ex)^2 = (ae - x)^2 + y^2$$

or

$$a^2 - 2aex + e^2x^2 = a^2e^2 - 2aex + x^2 + y^2$$

We can write this as

$$(1-e^2)x^2 + y^2 = a^2(1-e^2) = b^2$$

where

$$e = \sqrt{(1 - b^2/a^2)}$$

Once again, this confirms that $x^2/a^2 + y^2/b^2 = 1$. Expressing the equation of a circle in the form

$$x = r \cos \theta, \; y = r \sin \theta$$

or of an ellipse as

$$x = a \cos \theta, \; y = b \sin \theta$$

means that we have only one variable and can deduce both x and y from that variable. Lines and curves expressed in this way are said to be in *parametric form*, the parameter here being θ. The parabola $y^2 = 4ax$ can be expressed in parametric form as

$$x = at^2 \text{ and } y = 2at$$

There is then only one independent variable (t) and two dependent variables (x and y).

The parametric version of a straight line takes the form

$$x = p + lt, \; y = q + mt$$

where 'p', 'q', 'l', and 'm' are all constants and t is the variable (which in effect is the distance along the line from the point (p, q)), 'l' being the equivalent of cos (slope) and 'm' of sin (slope).

The parametric form of the circle with center (x_c, y_c) and radius r is

$$x = xc + r \cos \theta, \quad y = yc + r \sin \theta$$

Hence

$$dx/d\theta = -r \sin \theta, \; dy/d\theta = r \cos \theta$$

$$(dy/d\theta)/(dx/d\theta) = -(\cos \theta / \sin \theta) = -\cot \theta$$

or

$$dy/dx = -\cot \theta$$

This is the measure of the slope of the curve at the point θ. (Box 8.1)

Box 8.1 — Parametric Equations with 't' as a Variable

For a line: $\quad x = p + lt, \quad y = q + mt$

For a circle: $\quad x = p + r \cos t, \quad y = q + r \sin t$

For an ellipse: $\quad x = p + a \cos t, \quad y = q + b \sin t$

CURVES AND SURFACES

8.2 THE ELLIPSE

Figure 8.4 shows the auxiliary circle of radius 'a' and the ellipse that has a semi-minor axis of length 'b.' A point P on the ellipse has coordinates ($a \cos \theta, b \sin \theta$) referred to the center. PT is the tangent to the ellipse at point P, with T being on the major axis. PQ is the normal at P, with Q being on the minor axis. The line QP makes an angle ϕ with the major axis and is known as the *geodetic* or *spheroidal latitude* of P.

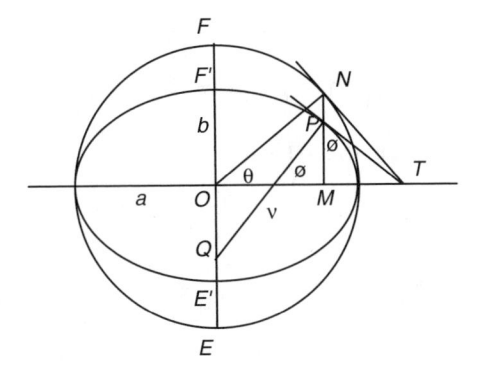

Figure 8.4 Normals to an ellipse.

From the parametric form of a circle, we can calculate the slope of the curve at any point, which is in fact the slope of the tangent at that point P that is, NT in Figure 8.4. Similarly, for the ellipse, the tangent is PT, where

$$dy/dx = (dy/d\theta)/(dx/d\theta) = (b \cos \theta)/(-a \sin \theta)$$
$$= -\frac{b}{a} \cot \theta$$

Circles and ellipses are important figures in geomatics since when rotated about their north/south axis, they create volumes that represent good approximations to the shape of the Earth, depending on their scale.

The angle POT (not shown in the diagram) would be the geocentric latitude but is not normally used. The angle θ is sometimes known as the *reduced latitude*. The length PQ is usually denoted by the Greek letter ν (pronounced 'nu' as in "new"). On an ellipsoid that is obtained by rotating the ellipse about its minor axis, the length of ν is the same for all points along a given parallel of latitude but has different values at different latitudes. The x coordinate of $P = a \cos \theta$ from before, but also is equal to $\nu \cos \phi$ = OM from the projection of QP onto the x-axis.

Hence $\qquad a \cos \theta = \nu \cos \phi.$

TN is tangential to the auxiliary circle since by squashing the auxiliary circle down to the ellipse, tangency will not be affected even though angles will change. Angle $MNT = \theta$, and angle $MPT = \phi$, $NM = a \sin \theta$, $OM = x$ value of $P = a \cos \theta$, and $PM = y$ value of $P = b \sin \theta$.

120 INTRODUCTION TO MATHEMATICAL TECHNIQUES USED IN GIS

Box 8.2 — The Radius v

From Figure 8.5

$$HI = a \cos \phi, \quad IJ = b \sin \phi, \quad \angle JHI = \theta$$

Thus $IJ/HI = \tan \theta = (b/a) \tan \phi$. Using Pythagoras, $HJ^2 = a^2 \cos^2 \phi + b^2 \sin^2 \phi$. But since $\cos^2 \phi + \sin^2 \phi = 1$ we can write this as

$$HJ^2 = a^2 - a^2 \sin^2 \phi + b^2 \sin^2 \phi = a^2 \left(1 - \frac{(a^2 - b^2)}{a^2} \sin^2 \phi\right)$$

$$= a^2 (1 - e^2 \sin^2 \phi)$$

where $e^2 = (1 - b^2/a^2)$ as before. Hence

$$HJ = a\sqrt{(1 - e^2 \sin^2 \phi)}$$

So in triangle HIJ,

$$\cos \theta = HI/HJ = a \cos \phi / (a \sqrt{(1 - e^2 \sin^2 \phi)})$$

or

$$a \cos \theta = a \cos \phi / \sqrt{(1 - e^2 \sin^2 \phi)}$$

We have previously shown that $a \cos \theta = v \cos \phi$. So $a \cos \theta = v \cos \phi = a \cos \phi / \sqrt{(1 - e^2 \sin^2 \phi)}$. Thus $v = a/\sqrt{(1 - e^2 \sin^2 \phi)}$ or

$$a = v \sqrt{(1 - e^2 \sin^2 \phi)}$$

But $x^2/a^2 + y^2/b^2 = 1$ and $x = v \cos \phi$.
So $y^2 = b^2 - (b^2/a^2)x^2 = a^2(1 - e^2) - (1 - e^2)x^2 = (1 - e^2)(a^2 - x^2)$.
Hence, since $a^2 = v^2(1 - e^2 \sin^2 \phi)$ and $x^2 = v^2 \cos^2 \phi = v^2(1 - \sin^2 \phi)$.
Thus, $(a^2 - x^2) = v^2(1 - e^2)\sin^2 \phi$, which results in

$$y^2 = v^2 (1 - e^2)^2 \sin^2 \phi$$

Thus

$$x = v \cos \phi \quad \text{and} \quad y = v(1 - e^2) \sin \phi \quad \text{and} \quad v = a/\sqrt{(1 - e^2 \sin^2 \phi)}$$

In triangle MNT, $MT = NM \tan \theta = a \sin \theta \tan \theta$. In triangle MPT, $MT = PM \tan \phi = b \sin \theta \tan \phi$. Hence, $a \tan \theta = b \tan \phi$. We can express this as

$$\tan \theta = (b \sin \phi)/(a \cos \phi)$$

Since angles are just ratios, we can represent this relationship by the right-angled triangle HIJ in Figure 8.5 in which $IJ = b \sin \phi$ and $HI = a \cos \phi$ and $\angle JHI = \theta$. Then by suitable manipulation as shown in Box 8.2, we can show that

$$x = v \cos \phi, \, y = v(1 - e^2) \sin \phi, \, v = a/\sqrt{(1 - e^2 \sin^2 \phi)}$$

CURVES AND SURFACES

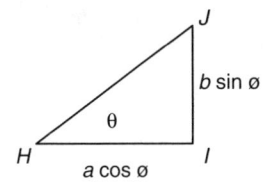

Figure 8.5 θ and ø.

8.3 THE RADIUS OF CURVATURE

Before leaving the ellipse, there is one further quantity that is required. It is known as the radius of curvature at P along the line of the ellipse and is normally referred to as 'ρ,' the Greek letter "rho" (pronounced row as in a row of beans). *Curvature* is the rate of change of the direction of a tangent in relation to the length of the arc; the *radius of curvature* at P is the radius of a circle with curvature equal to that of the curve at P, or, in other words, the radius of the circle that just touches the curve. At the point of contact, the tangent to the circle is also the tangent to the curve.

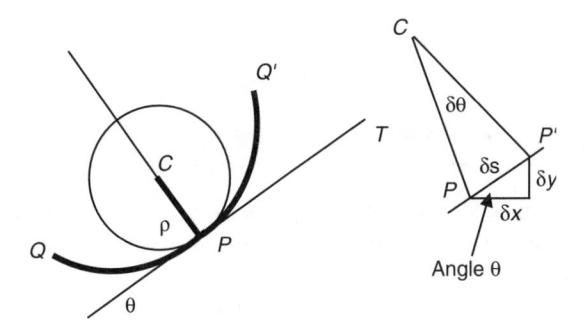

Figure 8.6 Radius of curvature.

Consider the curve QPQ' in Figure 8.6 with the circle center C just touching the curve at P. At the point of contact, the radius CP ($=\rho$) will be normal to the tangent PT that has a slope here called θ. Also consider the two nearby points P and P' shown in the inset. Their coordinates will differ by δx and δy, while the length along the curve is denoted by δs. If the slope at P is θ and at P' is θ + δθ, then the angle $PCP' = $ δθ (exaggerated in Figure 8.6).

By measuring the angle θ in radians in triangle PCP', δs = CPδθ = ρ δθ or (δθ/δs) = 1/ρ. The slope of the line $PP' = $ θ. Therefore, δy/δx = tan θ. The second derivative or curvature is

$$d^2y/dx^2 = d(\tan\theta)/dx$$

$$= \{d\theta/dx\} * \{d(\tan\theta)/d\theta\} = (d\theta/dx)\sec^2\theta$$

since the differential of tan θ = sec^2 θ. Also, $\delta\theta/\delta x = (\delta\theta/\delta s)*(\delta s/\delta x)$ and $\delta s/\delta x =$ sec θ as can be seen from Figure 8.6. So

$$d^2y/dx^2 = (1/\rho)\sec^3\theta$$

Sec^2 = 1 + tan^2. Also, tan θ = (dy/dx) from the small triangle in Figure 8.6. Therefore,

$$d^2y/dx^2 = (1/\rho)(1 + (dy/dx)^2)^{3/2}$$

On rearranging, the radius of curvature of any curve is given by

$$\rho = \{(1 + (dy/dx)^2)^{3/2}\}/(d^2y/dx^2)$$

If all this is applied to an ellipse, then it can be shown that

$$\rho = v(1-e^2)(1-e^2\sin^2\phi)^{-1} = a(1-e^2)(1-e^2\sin^2\phi)^{-3/2}$$

Box 8.3 — Radii of Curvature

For any curve, the radius of curvature

$$\rho = \{(1 + (dy/dx)^2)^{3/2}\}/(d^2y/dx^2)$$

For an ellipse

$$\rho = v(1-e^2)(1-e^2\sin^2\phi)^{-1}$$
$$= a(1-e^2)(1-e^2\sin^2\phi)^{-3/2}$$

$x = v\cos\phi$, $y = v(1-e^2)\sin\phi$, $v = a/\sqrt{(1-e^2\sin^2\phi)}$

The value 'ρ' (Box 8.3) is extensively used in geodetic computations; but these are beyond the scope of the present text. For now, we shall consider more about curvature. Every second-degree (quadratic) curve bends only one way. The circle and ellipse, for example, bend all the way round until they close back on themselves, while the parabola and hyperbola go off to infinity before turning back.

From Figure 8.6, we can also note that $\delta s^2 = \delta x^2 + \delta y^2$ or $\delta s/\delta x = \sqrt{1 + (\delta y/\delta x)^2}$. From this, we obtain $ds/dx = \sqrt{1 + (dy/dx)^2}$ or

$$s = \int\sqrt{1 + (dy/dx)^2}\,dx$$

This provides a method for calculating the length along a curve. In the particular case of the ellipse, the integration is not straightforward; but for some functions, it provides an elegant way to determine such lengths.

8.4 FITTING CURVES TO POINTS

The simplest curve that bends forwards and backwards is the cubic. If we take the general case of the cubic

$$y = a + bx + cx^2 + dx^3$$

CURVES AND SURFACES

Then,

$$dy/dx = b + 2cx + 3dx^2$$

There can be two values for which $dy/dx = 0$. Also,

$$d^2y/dx^2 = 2c + 6dx$$

There is only one value for which $d^2y/dx^2 = 0$ (see Figure 6.3). Because there are four unknown constants (a, b, c, d) in the equation for y, we need four independent equations to calculate them. In other words, a cubic can be made to pass through four points just as a quadratic curve such as a circle can be made to pass through any three points (albeit with infinite radius for three points on a straight line). Five points are needed to define a quartic (fourth degree), six for a quintic (five degrees), etc.

Given a cubic in the form $y = f(x)$ with four known points (x_1, y_1), (x_2, y_2), (x_3, y_3), and (x_4, y_4), the equation

$$y = y_1 \frac{(x - x_2)(x - x_3)(x - x_4)}{(x_1 - x_2)(x_1 - x_3)(x_1 - x_4)} + y_2 \frac{(x - x_1)(x - x_3)(x - x_4)}{(x_2 - x_1)(x_2 - x_3)(x_2 - x_4)}$$

$$+ y_3 \frac{(x - x_1)(x - x_2)(x - x_4)}{(x_3 - x_1)(x_3 - x_2)(x_3 - x_4)} + y_4 \frac{(x - x_1)(x - x_2)(x - x_3)}{(x_4 - x_1)(x_4 - x_2)(x_4 - x_3)}$$

is a cubic in 'x' and passes through all four points. Note that the sequence can be extended to quartics, quintics, and any other order and can be expressed in mathematical shorthand as

$$y = \sum_{i=1}^{n} y_i \prod_{i \neq j} \frac{(x - x_j)}{(x_i - x_j)}$$

where $\sum_{i=1}^{n}$ means the sum of all the y_i from $i = 1$ to n and $\prod_{i \neq j}$ means the product of the following expression, subject to the constraint that i does not equal j ($i \neq j$).

But what if we want to fit a cubic to five points? One solution would be to make the best-fit approximation to all five. Another would be to ensure that the curve goes exactly through all five points; but this is normally not possible with only one cubic. The solution is to fit the curve step by step in sections or what is called "*piecewise.*" This means fitting several cubic curves so that if we have the points A, B, C, D, E, ..., N then we fit the curve to AB, then BC, then CD, then DE, etc., until the last point N is reached. The result is what is called a *spline* curve, the points A, B, C, etc., being known as *nodes* or *knots*.

Before considering piecewise polynomials, let us return to the way in which we express a straight line between two points in parametric form:

$$x = a + bt, \quad y = c + dt$$

For a quadratic,

$$x = a + bt + ct^2, \quad y = d + et + ft^2$$

For a cubic,

$$x = a + bt + ct^2 + dt^3, \quad y = e + ft + gt^2 + ht^3$$

The reason for expressing (x, y) in this form is that with a computer plotter we can increment 't' from 0 to 1 in, for example, 10 steps of 0.1. The computer calculates (x, y) for $t = 0$, then for $t = 0.1$, then for $t = 0.2$, etc., until $t = 0.9$, and then $t = 1.0$. We would then have eleven values for (x, y) from the start point (x_s, y_s) where $t = 0$ to the end point (x_e, y_e), where $t = 1$. A computer-driven plotter can draw a series of straight lines joining these points consecutively so as to portray the curve from its beginning to its end. By taking finer increments of t (e.g., every 0.01 or 0.05 rather than 0.1), we can create the appearance of the smoother curve but at the cost of a longer processing time.

If we have two points A (x_1, y_1) and B (x_2, y_2), then any point on the line AB can be expressed in the form

$$x = x_1 + (x_2 - x_1) t \quad \text{and} \quad y = y_1 + (y_2 - y_1) t$$

(where $t = 0$ at the point A and $t = 1$ at B).

Consider three points A (x_A, y_A), B (x_B, y_B), and C (x_C, y_C) in Figure 8.7. A second-degree curve can be made to pass exactly through three points; but here we shall divide it into two separate but continuous sections. For the each section, AB and BC there is a quadratic of the form

$$x = a + bt + ct^2, \quad dx/dt = b + 2ct, \quad d^2x/dt^2 = 2c$$
$$y = e + ft + gt^2, \quad dy/dt = f + 2gt, \quad d^2y/dt^2 = 2g$$

Let us consider two sets of equations for the two sections of the curve from A to B and from B to C. Let any point on the first section of the curve be (x_1, y_1) and on the second section (x_2, y_2). These coordinates must satisfy the quadratic expressions:

$$x_1 = a_1 + b_1 t_1 + c_1 t_1^2, \quad dx/dt = b_1 + 2c_1 t_1, \quad d^2x/dt^2 = 2c_1$$
$$y_1 = e_1 + f_1 t_1 + g_1 t_1^2, \quad dy/dt = f_1 + 2g_1 t_1, \quad d^2y/dt^2 = 2g_1$$
$$x_2 = a_2 + b_2 t_2 + c_2 t_2^2, \quad dx/dt = b_2 + 2c_2 t_2, \quad d^2x/dt^2 = 2c_2$$
$$y_2 = e_2 + f_2 t_2 + g_2 t_2^2, \quad dy/dt = f_2 + 2g_2 t_2, \quad d^2y/dt^2 = 2g_2$$

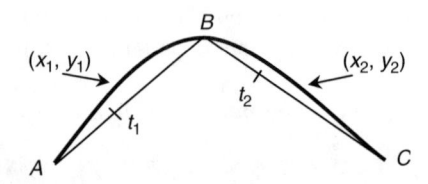

Figure 8.7 Fitting a second-degree curve.

CURVES AND SURFACES

For section 1 from A to B: At A, $t_1 = 0$ and $x_1 = a_1$ and $y_1 = e_1$. Thus $a_1 = x_A$ and $e_1 = y_A$. At B, $t_1 = 1$ and $x_1 = x_B = a_1 + b_1 + c_1$ or $b_1 + c_1 = x_B - x_A$. Likewise, $f_1 + g_1 = y_B - y_A$.

For section 2 from B to C: At B, $t_2 = 0$ and $x_2 = a_2$ and $y_2 = e_2$. Since this is the point B, $a_2 = x_B$ and $e_2 = y_B$. At C, $t_2 = 1$ and $x_2 = x_C = a_2 + b_2 + c_2$ or $b_2 + c_2 = x_C - x_B$. Likewise, $f_2 + g_2 = y_C - y_B$.

For the curve to be continuous at point B, the first and second derivatives must be the same. For the first derivative, this means that $b_1 + 2c_1 = b_2$, and $f_1 + 2g_1 = f_2$, while for the second derivative, $c_1 = c_2$ and $g_1 = g_2$. Putting all this together,

$$a_1 = x_A, \; a_2 = x_B, \; b_1 = (4x_B - 3x_A - x_C)/2$$

$$b_2 = (x_C - x_A)/2, \; c_1 = c_2 = (x_A - 2x_B + x_C)/2$$

Similarly,

$$e_1 = y_A, \; e_2 = y_B, \; f_1 = (4y_B - 3y_A - y_C)/2$$

$$f_2 = (y_C - y_A)/2, \; g_1 = g_2 = (y_A - 2y_B + y_C)/2$$

Table 8.1 Data for Points on a Piece-wise Quadratic

t	0	0.1	0.2	0.3	0.4	0.5	0.6	0.7	0.8	0.9	1
x_1	3.0	5.1	7.2	9.3	11.4	13.4	15.5	17.3	19.2	21.1	23.0
y_1	5.0	8.4	11.4	14.2	16.6	18.8	20.6	22.2	23.4	24.4	25
x_2	23.0	24.8	26.6	28.4	30.2	31.9	33.6	35.2	36.8	38.4	40
y_2	25	25.3	25.4	25.2	24.6	23.8	22.6	21.2	19.4	17.4	15

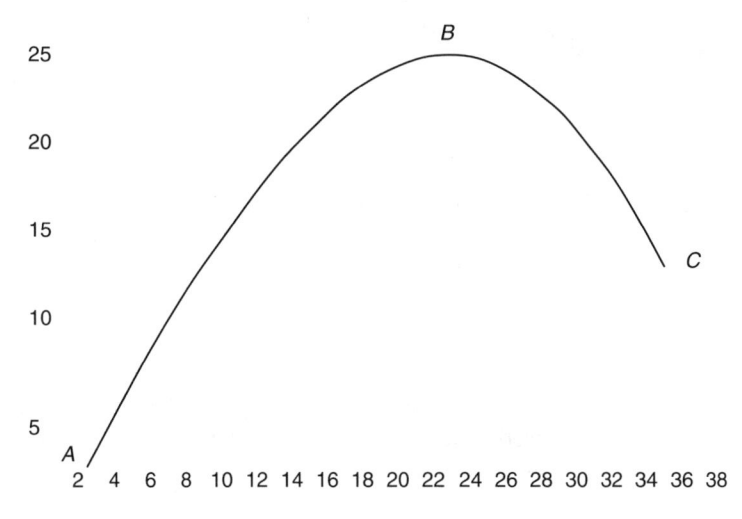

Figure 8.8 Two quadratics fitted to three points.

As an example of a piecewise quadratic, consider the points $A(3, 5)$, $B(23, 25)$, and $C(40,15)$ shown in Table 8.1 and Figure 8.8. Using the equations derived above,

$$a_1 = 3; \quad a_2 = 23; \quad b_1 = 21.5$$
$$b_2 = 18.5; \quad c_1 = c_2 = -1.5$$

Similarly,

$$e_1 = 5; \quad e_2 = 25; \quad f_1 = 35$$
$$f_2 = 5; \quad g_1 = g_2 = -15$$

The trouble with a quadratic is that it bends only one way — it cannot twist left and then right, clockwise then counterclockwise, or *vice versa*. A cubic can do this since it can have a maximum and minimum and a point of inflection in between. As a result, we can fit a set of cubic curves piecewise through any number of points.

Consider the series of points $A, B, C, D, ..., M, N$ in Figure 8.9. The upper part of Figure 8.9 shows a series of points joined by lines. Let these be $A(x_0, y_0)$, $B(x_1, y_1)$, $C(x_2, y_2)$, $D(x_3, y_3)$, ..., $M(x_{n-1}, y_{n-1})$, $N(x_n, y_n)$. The series of lines A–B–C, etc., is said to form a *string*.

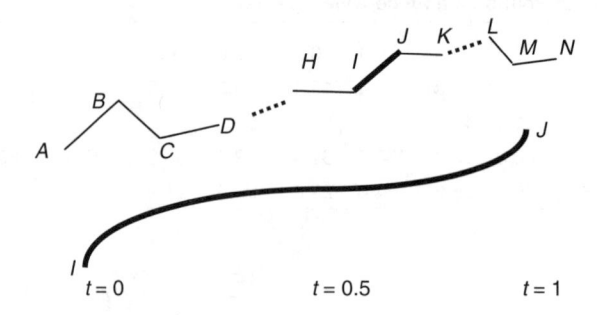

Figure 8.9 A piecewise cubic.

Now consider the two successive points along this string, $I(x_i, y_i)$ and $J(x_{i+1}, y_{i+1})$. For a cubic curve between I and J, we can let

$$x = a_i + b_i t + c_i t^2 + d_i t^3$$
$$y = e_i + f_i t + g_i t^2 + h_i t^3$$

where 't' is made to take a series of values from 0 to 1. We then have 8 unknowns ($a_i\ b_i\ c_i\ d_i\ e_i f_i\ g_i\ h_i$) for the section between i and j. But when $t = 0$, we are at I; hence $x_i = a_i$ and $y_i = e_i$. When $t = 1$, we are at J; hence

$$x_j = a_j + b_j + c_j + d_j \quad \text{and} \quad y_j = e_j + f_j + g_j + h_j$$

CURVES AND SURFACES

or

$$x_j - x_i = b_j + c_j + d_j \quad \text{and} \quad y_j - y_i = f_j + g_j + h_j$$

Also,

$$dx/dt = b_i + 2c_i t + 3d_i t^2, \quad dy/dt = f_i + 2g_i t + 3h_i t^2$$

$$d^2x/dt^2 = 2c_i + 6d_i t, \quad d^2y/dt^2 = 2g_i + 6h_i t$$

To obtain a smooth curve at I, there must be continuity in both the first and second differentials so that both the slope and rate of curvature are continuous, which means that at I (where $t = 0$), $dx/dt = b_i$ and must have the same value as the previous section HI where $t = 1$. This would have been $b_h + 2c_h + 3d_h$ using the same notation. Hence, $b_i = b_h + 2c_h + 3d_h$ and similarly $f_i = f_h + 2g_h + 3h_h$.

At point I on the section IJ, where $t = 0$, $d^2x/dt^2 = 2c_i$ and this must equal the value from the previous section of the curve when $t = 1$ namely $2c_h + 6d_h$. Hence, we can calculate c_i and similarly g_i and since we have

$$d_i = (x_j - x_i) - b_j - c_j \quad \text{and} \quad h_i = (y_j - y_i) - f_j - g_j$$

we have all the parameters for the section IJ, provided that we have established the parameters for HI. Thus, for all the sections of the curve BC, CD, ..., IJ, ..., MN, we can fit a cubic in parametric form, based on values obtained from the previous section. The trouble is how to get started, since we need to know the initial slope and curvature in order to calculate the section AB. There is an infinite number of possibilities.

Of the many practical solutions, the most common is to set the second derivative to zero at the very start and very end of the string. An alternative is to use the values from the quadratic described above as the start for the line AB and carry through the values at B as the initial values for fitting a cubic to BC, etc.

There is, however, a fundamental problem with the approach described in that although it works under many circumstances, it can lead to a curve that crosses over itself as in Figure 8.10.

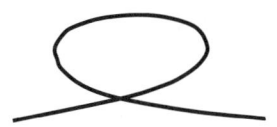

Figure 8.10 Looping curve.

Such a loop is unacceptable if it is meant to represent a contour line through a set of interpolated points. It is not the intention of this text to discuss curve-fitting

8.5 THE BEZIER CURVE

There is, however, one other approach that will be mentioned, since it is commonly referred to in computer drawing packages. Known as the family of Bezier polynomials, they take the form

$$x = f_x(t) = \sum_{r=0}^{n} {_nC_r} t^i (1-t)^{n-i} x_i$$

$$y = f_y(t) = \sum_{r=0}^{n} {_nC_r} t^i (1-t)^{n-i} y_i$$

where 'f' means the function for which 't' takes a series of steps from 0 to 1 calculated separately for each x_i and y_i and

$$_nC_r = n!/\{r!(n-r)!\}$$

The curve starts at A (x_o, y_o) and finishes at B (x_n, y_n) with $(n-1)$ guiding points P_1 (x_1, y_1), P_2 (x_2, y_2) through to $P_{(n-1)}$ (x_{n-1}, y_{n-1}). A and B can be regarded as P_0 and P_n.

The curve starts tangential to the line P_0P_1 (AP_1 in Figure 8.11) and finishes tangential to the line $P_{n-1}P_n$ (here P_2B). For any point 't' along a Bezier curve of the third degree, we have

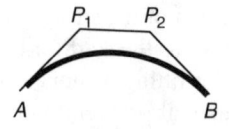

Figure 8.11 A Bezier curve with two control points.

$$x_t = (1-t)^3 x_0 + 3t(1-t)^2 x_1 + 3t^2(1-t)x_2 + t^3 x_3$$

$$y_t = (1-t)^3 y_0 + 3t(1-t)^2 y_1 + 3t^2(1-t)y_2 + t^3 y_3$$

If A (20, 20) is to be joined to B (40, 25) with P_1 (25, 25) and P_2 (35, 35), then the curve would follow the upper bold line in Figure 8.12. If the y value of P_2 were changed to (35, 28), then the curve would follow the lower, thinner, line — calculated here as y', x remaining the same (see Table 8.2).

CURVES AND SURFACES

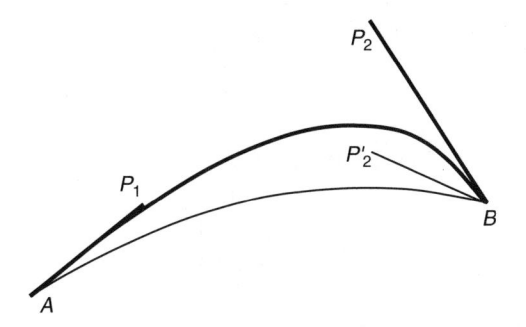

Figure 8.12 Two versions of a Bezier curve.

Table 8.2 Data for a Bezier Curve

t	1−t	(1−t)³	3t(1−t)²	3t²(1−t)	t³	x	y	y'
0	1	1	0	0	0	20	20	20
0.1	0.9	0.729	0.243	0.027	0.001	21.64	21.625	21.409
0.2	0.8	0.512	0.384	0.096	0.008	23.52	23.4	22.632
0.3	0.7	0.343	0.441	0.189	0.027	25.58	25.175	23.663
0.4	0.6	0.216	0.432	0.288	0.064	27.76	26.8	24.496
0.5	0.5	0.125	0.375	0.375	0.125	30	28.125	25.125
0.6	0.4	0.064	0.288	0.432	0.216	32.24	29	25.544
0.7	0.3	0.027	0.189	0.441	0.343	34.42	29.275	25.747
0.8	0.2	0.008	0.096	0.384	0.512	36.48	28.8	25.728
0.9	0.1	0.001	0.027	0.243	0.729	38.36	27.425	25.481
1	0	0	0	0	1	40	25	25

Thus, the selection of the control points P_1, P_2, etc., is critical to the shape of the curve. By choosing various controlling points, a graphic designer can adjust a curve to meet any required design. In general, in matrix notation, if

$$\mathbf{P}_i = \begin{bmatrix} x_i \\ y_i \end{bmatrix} \text{ and } \mathbf{P}(t) = \begin{bmatrix} p_x(t) \\ p_y(t) \end{bmatrix},$$

then

$$\mathbf{P}(t) = \sum_{r=0}^{r=n} {}_nC_r t^r (1-t)^{n-r} \mathbf{P}_r$$

The idea of curve fitting can be extended to three dimensions. Thus, if we are fitting the cubic in the form

$$x = a_i + b_i t + c_i t^2 + d_i t^3$$

and

$$y = e_i + f_i t + g_i t^2 + h_i t^3$$

then we can add

$$z = l_i + m_i t + n_i t^2 + o_i t^3$$

Similarly, the Bezier function can be extended so that

$$\mathbf{P}(t) = \begin{bmatrix} p_x(t) \\ p_y(t) \\ p_z(t) \end{bmatrix}$$

Curves and surfaces may be made to go exactly through a series of fixed points or can be constructed as the best mean fit. The latter requires a statistical approach and is discussed in Chapters 10 and 11.

CHAPTER 9

Transformations

CONTENTS

9.1 Homogeneous Coordinates 131
9.2 Rotating an Object..132
9.3 Hidden Lines and Surfaces 139
9.4 Map Projections...140
9.5 Cylindrical Projections 142
9.6 Azimuthal Projections 145
9.7 Conical Projections 147

9.1 HOMOGENEOUS COORDINATES

In this chapter, we will consider two forms of transformation, both essentially concerned with two-dimensional (2D) representation of three-dimensional (3D) objects. The first relates to what is sometimes called visualization and the display of the landscape in perspective while the second is more commonly referred to as map projections.

Firstly, however, we will extend the idea of 3D coordinates (x, y, z) to homogeneous coordinates (x, y, z, w) where in effect the 'w' is a scaling factor and not a fourth dimension. The actual 3D coordinates then are, $(x/w, y/w, z/w)$. In the 2D case, the homogeneous coordinates would be (x, y, w) with the traditional values with which we normally deal being $(x/w, y/w)$. The reason for introducing this extra complexity is to allow us to describe points that are at infinity.

Consider the line AB and a point C in Figure 9.1. Every line through C cuts the line AB somewhere (e.g., CD, CE, or CF); but when the line through C is parallel to AB it is an infinite distance away, although in a known direction. It is not just anywhere but specifically in the direction in which AB is pointing.

If in conventional terms AB is the line $y = mx + c$, then if we use homogeneous coordinates we can express the point at infinity as $(x, mx, 0)$ and still retain the relationship that 'y' is basically 'm' times 'x.' We can use this relationship to draw

131

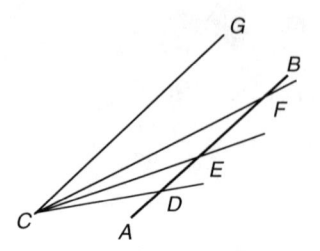

Figure 9.1 Points at infinity.

perspective, as parallel lines are made to appear to converge on the horizon as illustrated in Figure 9.2 at what are called the vanishing points.

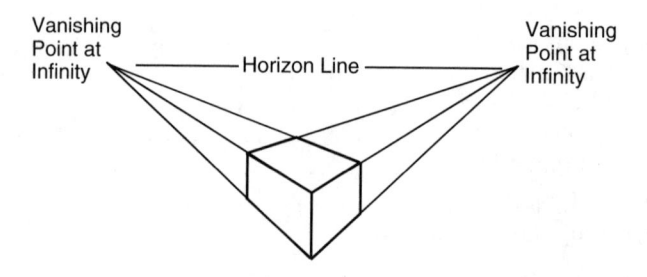

Figure 9.2 Vanishing points for a rectangular block.

9.2 ROTATING AN OBJECT

We will start by considering a 4∗4 matrix transformation of a point (x, y, z, w) of the form:

$$\begin{pmatrix} \cos\theta & -\sin\theta & 0 & 0 \\ \sin\theta & \cos\theta & 0 & 0 \\ 0 & 0 & 1 & 0 \\ 0 & 0 & 0 & 1 \end{pmatrix} * \begin{pmatrix} x \\ y \\ z \\ w \end{pmatrix} = \begin{pmatrix} x\cos\theta - y\sin\theta \\ x\sin\theta + y\cos\theta \\ z \\ w \end{pmatrix}$$

and this results in the new co-ordinates $\{(x\cos\theta - y\sin\theta), (x\sin\theta + y\cos\theta), z, w\}$. As we saw in Chapter 7, this represents a rotation about the z-axis and we can call the matrix $\mathbf{R}_z$. Similarly, with respect to the y-axis, we have

$$\mathbf{R}_y = \begin{pmatrix} \cos\phi & 0 & -\sin\phi & 0 \\ 0 & 1 & 0 & 0 \\ \sin\phi & 0 & \cos\phi & 0 \\ 0 & 0 & 0 & 1 \end{pmatrix}$$

TRANSFORMATIONS

and for the rotation about the x-axis

$$\mathbf{R}_x = \begin{pmatrix} 1 & 0 & 0 & 0 \\ 0 & \cos\omega & -\sin\omega & 0 \\ 0 & \sin\omega & \cos\omega & 0 \\ 0 & 0 & 0 & 1 \end{pmatrix}$$

In Chapter 7, we called these two matrices $\mathbf{R}_y$ and $\mathbf{R}_x$ and discussed how we can combine them into one matrix:

$$\mathbf{R} = \begin{pmatrix} a & b & c & 0 \\ d & e & f & 0 \\ g & h & i & 0 \\ 0 & 0 & 0 & 1 \end{pmatrix}$$

The values of 'a', 'b', 'c,' etc. will depend on the sequence in which the rotations take place but so long as the original rotations represent orthogonal transformations, the effect will be to keep right angles as right angles. If R is not an orthogonal matrix, the axes will become skewed.

Consider next the transformation

$$\begin{pmatrix} 1 & 0 & 0 & t \\ 0 & 1 & 0 & u \\ 0 & 0 & 1 & v \\ 0 & 0 & 0 & 1 \end{pmatrix} * \begin{pmatrix} x \\ y \\ z \\ w \end{pmatrix} = \begin{pmatrix} x + tw \\ y + uw \\ z + vw \\ w \end{pmatrix}$$

resulting in new coordinates $(x', y', z') = (x/w + t, y/w + u, z/w + v)$. This is a simple translation of the origin from its original point to $(-t, -u, -v)$.

Now consider

$$\begin{pmatrix} 1 & 0 & 0 & 0 \\ 0 & 1 & 0 & 0 \\ 0 & 0 & 1 & 0 \\ 0 & 0 & 0 & s \end{pmatrix} * \begin{pmatrix} x \\ y \\ z \\ w \end{pmatrix}$$

This results in

$$\begin{pmatrix} x \\ y \\ z \\ sw \end{pmatrix}$$

or in the more conventional form,

$$x' = x/sw, \ y' = y/sw \quad \text{and} \quad z' = z/sw$$

This means that all dimensions have been reduced in scale by an amount = 1/s. Finally, consider

$$\begin{pmatrix} 1 & 0 & 0 & 0 \\ 0 & 1 & 0 & 0 \\ 0 & 0 & 1 & 0 \\ p & q & r & s \end{pmatrix} * \begin{pmatrix} x \\ y \\ z \\ w \end{pmatrix}$$

This results in

$$\begin{pmatrix} x \\ y \\ z \\ px + qy + rz + sw \end{pmatrix}$$

The expression $px + qy + rz + sw$ is the equation of a plane and all values will be reduced to a plane that is skew to our original axes. So taken overall, if we apply the matrix

$$\mathbf{M} = \begin{pmatrix} a & b & c & t \\ d & e & f & u \\ g & h & i & v \\ p & q & r & s \end{pmatrix}$$

to the set of coordinates $(x, y, z, w)^T$, then we will change the origin, rotate the axes, change the overall scale, and project the whole scene onto a plane. Thus, by suitable choice of the 16 elements, we can project any string of coordinates from three dimensions into two.

Consider a house with a ridged roof for which the coordinates of the floor are

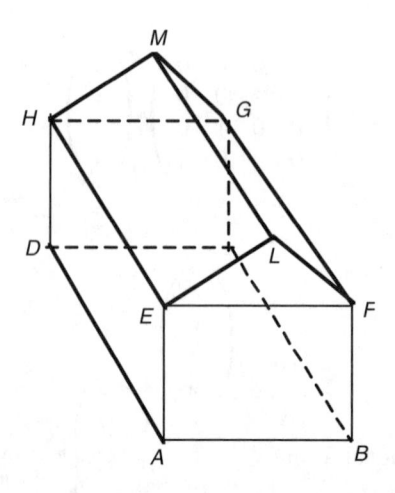

A (10, 20,100), B (20, 20,100), C (20,40, 100) and D (10, 40, 100)

Figure 9.3 A barn.

TRANSFORMATIONS

Let the line of the eaves be

$$E\ (10,\ 20, 108),\ F\ (20,\ 20, 108),\ G\ (20,\ 40,\ 108)\ \text{and}\ H\ (10,\ 40,\ 108)$$

and the ridge be $L\ (15,\ 20, 112),\ M\ (15,\ 40, 112)$.

The AB direction is x, the AD is y, and the AE is z. Let us express all these 10 points as homogeneous coordinates in a matrix **N** where the columns represent the coordinates of the corner points (Figure 9.3).

$$\mathbf{N} = \begin{pmatrix} A & B & C & D & E & F & G & H & L & M \\ 10 & 20 & 20 & 10 & 10 & 20 & 20 & 10 & 15 & 15 \\ 20 & 20 & 40 & 40 & 20 & 20 & 40 & 40 & 20 & 40 \\ 100 & 100 & 100 & 100 & 108 & 108 & 108 & 108 & 112 & 112 \\ 1 & 1 & 1 & 1 & 1 & 1 & 1 & 1 & 1 & 1 \end{pmatrix}$$

Let us first of all change the origin to the point A (10, 20, 100, 1) by applying the matrix **T** where

$$\mathbf{T} = \begin{pmatrix} 1 & 0 & 0 & -10 \\ 0 & 1 & 0 & -20 \\ 0 & 0 & 1 & -100 \\ 0 & 0 & 0 & 1 \end{pmatrix}$$

$$\mathbf{T} * \mathbf{N} = \begin{pmatrix} 1 & 0 & 0 & -10 \\ 0 & 1 & 0 & -20 \\ 0 & 0 & 1 & -100 \\ 0 & 0 & 0 & 1 \end{pmatrix} * \begin{pmatrix} 10 & 20 & 20 & 10 & 10 & 20 & 20 & 10 & 15 & 15 \\ 20 & 20 & 40 & 40 & 20 & 20 & 40 & 40 & 20 & 40 \\ 100 & 100 & 100 & 100 & 108 & 108 & 108 & 108 & 112 & 112 \\ 1 & 1 & 1 & 1 & 1 & 1 & 1 & 1 & 1 & 1 \end{pmatrix}$$

Call this

$$\mathbf{N}' = \begin{pmatrix} 0 & 10 & 10 & 0 & 0 & 10 & 10 & 0 & 5 & 5 \\ 0 & 0 & 20 & 20 & 0 & 0 & 20 & 20 & 0 & 20 \\ 0 & 0 & 0 & 0 & 8 & 8 & 8 & 8 & 12 & 12 \\ 1 & 1 & 1 & 1 & 1 & 1 & 1 & 1 & 1 & 1 \end{pmatrix}$$

This is a simple translation of axes by moving the location of the origin. Now let us rotate the building about the z-axis by 30° (with cos 30 = 0.866 and sin 30 = 0.5). The transformation achieving this is

$$\mathbf{R}_z * \mathbf{N}' = \begin{pmatrix} 0.866 & -0.5 & 0 & 0 \\ 0.5 & 0.866 & 0 & 0 \\ 0 & 0 & 1 & 0 \\ 0 & 0 & 0 & 1 \end{pmatrix} * \mathbf{N}' = \mathbf{N}'_z$$

INTRODUCTION TO MATHEMATICAL TECHNIQUES USED IN GIS

Our ten points then become

$$
\mathbf{N'}_z = \begin{pmatrix} & A & B & C & D & E & F & G & H & L & M \\ & 0 & 8.66 & -1.34 & -10 & 0 & 8.66 & -1.34 & -10 & 4.33 & -5.67 \\ & 0 & 5 & 22.32 & 17.32 & 0 & 5 & 22.32 & 17.32 & 2.5 & 19.82 \\ & 0 & 0 & 0 & 0 & 8 & 8 & 8 & 8 & 12 & 12 \\ & 1 & 1 & 1 & 1 & 1 & 1 & 1 & 1 & 1 & 1 \end{pmatrix}
$$

Thus, A has become $(0, 0, 0, 1)$, B $(8.66, 5, 0, 1)$, etc. Now let us tip it all backward by rotating the shape by $10°$ about the x-axis (clockwise for right-handed systems; $\cos 10 = 0.985$ and $\sin 10 = 0.174$). The transformation matrix then is

$$
\mathbf{R}_x * \mathbf{N'}_z = \begin{pmatrix} 1 & 0 & 0 & 0 \\ 0 & 0.985 & 0.174 & 0 \\ 0 & -0.174 & 0.985 & 0 \\ 0 & 0 & 0 & 1 \end{pmatrix} * \mathbf{N'}_z = \mathbf{N'}_{xz}
$$

where

$$
\mathbf{N'}_{xz} = \begin{pmatrix} 0 & 8.66 & -1.34 & -10 & 0 & 8.66 & -1.34 & -10 & 4.33 & -5.67 \\ 0 & 4.92 & 21.99 & 17.06 & 1.39 & 6.31 & 23.38 & 18.45 & 4.55 & 21.61 \\ 0 & -0.87 & -3.88 & -3.01 & 7.88 & 7.01 & 4.00 & 4.87 & 11.38 & 8.37 \\ 1 & 1 & 1 & 1 & 1 & 1 & 1 & 1 & 1 & 1 \end{pmatrix}
$$

Note that we could have combined all these three operations into one matrix $\mathbf{R}$:

$$
\mathbf{R} = \begin{pmatrix} 1 & 0 & 0 & 0 \\ 0 & 0.985 & 0.174 & 0 \\ 0 & -0.174 & 0.985 & 0 \\ 0 & 0 & 0 & 1 \end{pmatrix} * \begin{pmatrix} 0.866 & -0.5 & 0 & 0 \\ 0.5 & 0.866 & 0 & 0 \\ 0 & 0 & 1 & 0 \\ 0 & 0 & 0 & 1 \end{pmatrix} * \begin{pmatrix} 1 & 0 & 0 & -10 \\ 0 & 1 & 0 & -20 \\ 0 & 0 & 1 & -100 \\ 0 & 0 & 0 & 1 \end{pmatrix}
$$

$$
= \begin{pmatrix} 0.866 & -0.5 & 1 & 1.34 \\ 0.492 & 0.853 & 0.174 & -39.38 \\ -0.087 & -0.151 & 0.985 & -94.62 \\ 0 & 0 & 0 & 1 \end{pmatrix}
$$

Note

$$
\mathbf{R} * \mathbf{N} = \begin{pmatrix} 0.866 & -0.5 & 1 & 1.34 \\ 0.492 & 0.853 & 0.174 & -39.38 \\ -0.087 & -0.151 & 0.985 & -94.62 \\ 0 & 0 & 0 & 1 \end{pmatrix}
$$

$$
* \begin{pmatrix} 10 & 20 & 20 & 10 & 10 & 20 & 20 & 10 & 15 & 15 \\ 20 & 20 & 40 & 40 & 20 & 20 & 40 & 40 & 20 & 40 \\ 100 & 100 & 100 & 100 & 108 & 108 & 108 & 108 & 112 & 112 \\ 1 & 1 & 1 & 1 & 1 & 1 & 1 & 1 & 1 & 1 \end{pmatrix}
$$

$$
= \mathbf{N'}_{xz}
$$

TRANSFORMATIONS

This N'_{xz} is the one derived above. Note also that the sequence is important. If we had carried out the rotation about the x-axis before carrying out the rotation about the z, we would have obtained a different answer:

$$\mathbf{R'} = \begin{pmatrix} 0.866 & -0.5 & 0 & 0 \\ 0.5 & 0.866 & 0 & 0 \\ 0 & 0 & 1 & 0 \\ 0 & 0 & 0 & 1 \end{pmatrix} * \begin{pmatrix} 1 & 0 & 0 & 0 \\ 0 & 0.985 & 0.174 & 0 \\ 0 & -0.174 & 0.985 & 0 \\ 0 & 0 & 0 & 1 \end{pmatrix} * \begin{pmatrix} 1 & 0 & 0 & -10 \\ 0 & 1 & 0 & -20 \\ 0 & 0 & 1 & -100 \\ 0 & 0 & 0 & 1 \end{pmatrix}$$

$$= \begin{pmatrix} 0.866 & -0.492 & -0.082 & 9.89 \\ 0.5 & 0.853 & 0.151 & -37.17 \\ 0 & -0.174 & 0.985 & -95.02 \\ 0 & 0 & 0 & 1 \end{pmatrix}$$

which is not quite the same as the original **R**. Likewise, if the origin is moved after rather than before the rotations, there will be a different solution.

The matrix $\mathbf{N'}_{xz}$ represents the coordinates in space of the ten points after the rotation of the axes. There is no change in shape and no change in scale. The lengths of the sides are the same as before and right angles remain right angles in 3D space. This is a *similarity transformation* in that it is a simple rotation and translation in which orthogonal lines remain orthogonal. A similarity transformation can also incorporate a uniform scale change. It is also *affine* in that it preserves collinearity and parallelism, straight lines remaining straight lines.

The image shown in Figure 9.4 plots the x-axis from left to right, the z-axis from the bottom to the top of the page while the y values have been treated as zero and the image displayed on the plane $y = 0$.

In Figure 9.4, the 3D coordinates show that the lengths *DH*, *AE*, and *FB* are all the same as are *DA*, *HE*, and *ML*, etc. *EF* and *AB* are parallel and meet at a point on the x-axis that is an infinite distance away. This point can be given the coordinates (1, 0, 0, 0). If we multiply this by the skew matrix:

$$\mathbf{S}_x = \begin{pmatrix} 1 & 0 & 0 & 0 \\ 0 & 1 & 0 & 0 \\ 0 & 0 & 1 & 0 \\ 0.025 & 0 & 0 & 1 \end{pmatrix} \text{ then } \mathbf{S}_x * \begin{pmatrix} 1 \\ 0 \\ 0 \\ 0 \end{pmatrix} = \begin{pmatrix} 1 \\ 0 \\ 0 \\ 0.025 \end{pmatrix}$$

and the point at infinity becomes the point (1, 0, 0, 0.025) or in conventional coordinates (40, 0, 0). Similarly, points in the y-direction can be drawn to the point at infinity in the y-direction (0, 1, 0, 0). Thus

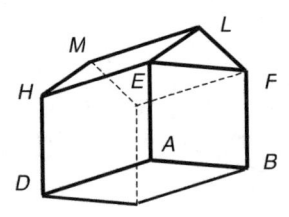

Figure 9.4 The barn after two rotations.

$$S_{xy} = \begin{pmatrix} 1 & 0 & 0 & 0 \\ 0 & 1 & 0 & 0 \\ 0 & 0 & 1 & 0 \\ 0.025 & 0.025 & 0 & 1 \end{pmatrix}$$

will bring both the points at infinity in the x- and y- directions into a range that can be plotted on a sheet of paper. If we apply this to achieve a perspective view, (Figure 9.5), then

$$\begin{pmatrix} 1 & 0 & 0 & 0 \\ 0 & 1 & 0 & 0 \\ 0 & 0 & 1 & 0 \\ 0.025 & 0.025 & 0 & 1 \end{pmatrix}$$

$$* \begin{pmatrix} 0 & 8.66 & -1.34 & -10 & 0 & 8.66 & -1.34 & -10 & 4.33 & -5.67 \\ 0 & 4.92 & 21.99 & 17.06 & 1.39 & 6.31 & 23.38 & 18.45 & 4.55 & 21.61 \\ 0 & -0.87 & -3.88 & -3.01 & 7.88 & 7.01 & 4.00 & 4.87 & 11.38 & 8.37 \\ 1 & 1 & 1 & 1 & 1 & 1 & 1 & 1 & 1 & 1 \end{pmatrix}$$

$$= \begin{pmatrix} 0 & 8.66 & -1.34 & -10 & 0 & 8.66 & -1.34 & -10 & 4.33 & -5.67 \\ 0 & 4.92 & 21.99 & 17.06 & 1.39 & 6.31 & 23.38 & 18.45 & 4.55 & 21.61 \\ 0 & -0.87 & -3.88 & -3.01 & 7.88 & 7.01 & 4.00 & 4.87 & 11.38 & 8.37 \\ 1 & 1.34 & 1.52 & 1.17 & 1.03 & 1.37 & 1.55 & 1.21 & 1.22 & 1.40 \end{pmatrix}$$

$$= \begin{pmatrix} 0 & 6.46 & -0.88 & -8.55 & 0 & 6.32 & -0.86 & -8.26 & 3.55 & -4.05 \\ 0 & 3.67 & 14.47 & 14.58 & 1.35 & 4.61 & 15.08 & 15.25 & 3.73 & 15.43 \\ 0 & -0.65 & -2.55 & -2.57 & 7.65 & 5.12 & 2.58 & 4.02 & 9.33 & 5.98 \\ 1 & 1 & 1 & 1 & 1 & 1 & 1 & 1 & 1 & 1 \end{pmatrix}$$

Thus, in non-homogeneous terms, we have for the perspective view:

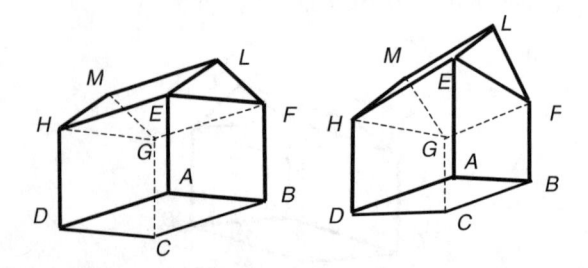

A (0, 0, 0), B (6.5, 3.7, −0.7), C (−0.9, 14.5, −2.6), D (−8.5, 14.6, −2.6)

E (0, 1.3, 7.6), F (6.3, 4.6, 5.1), G (−0.9, 15.1, 2.6), H (−8.3, 15.2, 4.0)

L (3.5, 2.7, 9.3), M (−4.1, 15.4, 6.0)

Figure 9.5 The affine and perspective projections of the barn.

TRANSFORMATIONS

Note that the two points at infinity (1, 0, 0, 0) and (0, 1, 0, 0) under the original transformation became

$$\begin{pmatrix} 0.866 & -0.5 & 0 & 1.34 \\ 0.492 & 0.853 & 0.174 & -39.38 \\ -0.087 & -0.151 & 0.985 & -94.62 \\ 0 & 0 & 0 & 1 \end{pmatrix} * \begin{pmatrix} 1 & 0 \\ 0 & 1 \\ 0 & 0 \\ 0 & 0 \end{pmatrix} = \begin{pmatrix} 0.866 & -0.5 \\ 0.492 & 0.853 \\ -0.087 & -0.151 \\ 0 & 0 \end{pmatrix}$$

and were still at infinity. However, by applying the matrix S_{xy} we obtain

$$\begin{pmatrix} 1 & 0 & 0 & 0 \\ 0 & 1 & 0 & 0 \\ 0 & 0 & 1 & 0 \\ 0.025 & 0.025 & 0 & 1 \end{pmatrix} * \begin{pmatrix} 0.866 & -0.5 \\ 0.492 & 0.853 \\ -0.087 & -0.151 \\ 0 & 0 \end{pmatrix} = \begin{pmatrix} 0.866 & -0.5 \\ 0.492 & 0.853 \\ -0.087 & -0.151 \\ 0.034 & 0.009 \end{pmatrix}$$

$$= \begin{pmatrix} 25.47 & -55.6 \\ 14.47 & 94.78 \\ -2.56 & -16.78 \\ 1 & 1 \end{pmatrix}$$

Thus, the lines *AB*, *EF*, *HG*, and *DC* converge at (25.5, 14.5, −2.6), while *AD*, *EH*, *FG*, *BC*, and *LM* converge on (−55.6, 94.8, −16.8). Transformations of the kind illustrated above can be used, for instance, to simulate a 3D landscape, including the effects of changing viewing positions, for instance, for "fly-by" presentations. Each point or line can be transformed and then displayed at its new coordinates. Straight lines remain straight but parallel lines can converge.

9.3 HIDDEN LINES AND SURFACES

In Chapter 3, we noted that for two points on a line, A (x_A, y_A) and B (x_B, y_B), we have

$$(y - y_A) = \frac{(y_B - y_A)}{(x_B - x_A)} * (x - x_A)$$

Given a point P (x_P, y_P), if we calculate

$$y = y_A + \frac{(y_B - y_A)}{(x_B - x_A)} * (x_P - x_A)$$

and if $(y_P > y)$ then P is above the line AB, while if $y_P < y$ then P lies below the line. We can extend this idea to any plane through the points *ABC*, which at the end of Chapter 7 we showed to have the form

$$(x - X_C)\{Y_A(Z_B - Z_C) + Y_B(Z_C - Z_A) + Y_C(Z_A - Z_B)\}$$
$$+ (y - Y_C)\{Z_A(X_B - X_C) + Z_B(X_C - X_A) + Z_C(X_A - X_B)\}$$
$$+ (z - Z_C)\{X_A(Y_B - Y_C) + X_B(Y_C - Y_A) + X_C(Y_A - Y_B)\}$$

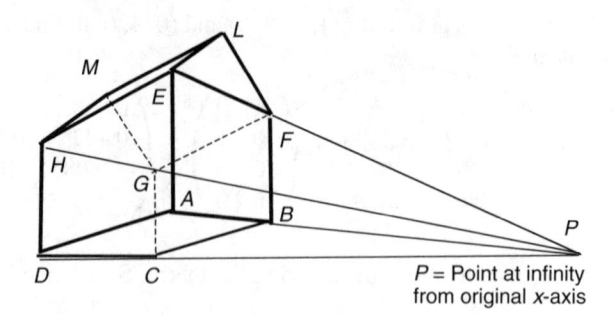

Figure 9.6 Transformation into a perspective view.

When we insert values for x, y, and z, this quantity will be either positive or negative unless the point lies on the plane. Two points on the same side of the plane will share the same sign (i.e., both will give a positive value or both will give a negative value).

We can use this fact to determine as to which lines are "hidden" when we view our solid object projected onto a flat surface. Every plane surface can be defined by three points; thus, for example, in Figure 9.6 for the surface defined by the four points A, D, E, and H we only need the coordinates of three of these. If we take a coordinate value inside the building (say a point Q at $(0, 5, 0)$ in the above example) and compare it with our viewing position (say somewhere along the y-axis at V $(0, -50, 0)$, then Q and V will be found to be on opposite sides of the plane ADE but on the same side of any of the planes that contains the point G, which is therefore invisible. Each surface will need to be tested, but this kind of routine calculation is ideally suitable for electronic data processing.

9.4 MAP PROJECTIONS

The assumption behind all the transformations considered so far has been that we have been dealing with straight lines or flat surfaces. When we wish to represent the curved surface of the Earth, we need to adopt a different approach.

Consider a point A on the Earth's surface with latitude $ø_A$ and longitude λ_A (Figure 9.7). The length of the arc from the equator (Q) to $A = R\, ø_A$, where R is the radius of the Earth, assumed here to be a sphere, and $ø$ is measured in radians. Let C be a point with latitude $ø_C$ and longitude λ_C. Let the difference in longitude between A and $C = \delta\lambda = \lambda_C - \lambda_A$ and the difference in latitude $\delta ø = ø_C - ø_A$. Let B be on the same parallel of latitude as C and the same meridian of longitude as A. In triangle ABC, then, in terms of physical length: $AB = R\delta ø$. Since the radius of the circle for the parallel of latitude $ø_C = PB$ and $PB = PC = R \cos ø_C$, then $BC = PB\, \delta\lambda = R \cos ø_C\, \delta\lambda$.

These quantities form the basis for plotting on a flat surface with a rectangular grid (Figure 9.8). AB on the sphere becomes $A'B' = \delta N$ and BC becomes $B'C' = \delta E$, the differences in Northings and Eastings between A' and C'.

TRANSFORMATIONS

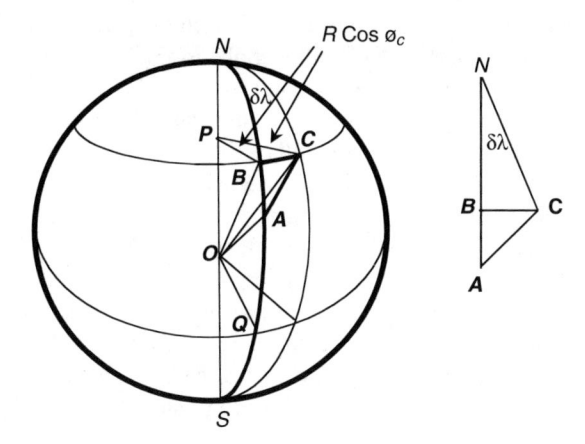

Figure 9.7 The elemental triangle.

In the simplest map projection, known as the "simple cylindrical" or "Plate Carree" projection, the distance north is plotted as $N = R\phi$ and the Eastings as $R\lambda$ so that there is a rectangular grid. This means, in effect, that the distances North–South are treated as correct but distances West–East must be stretched from $(R \cos \phi)\lambda$ on the globe to $R\lambda$ on the grid. East–West distances on the globe must be increased by a scale factor of $(1/\cos \phi)$ or $\sec \phi$ in order to plot them on the rectangular grid.

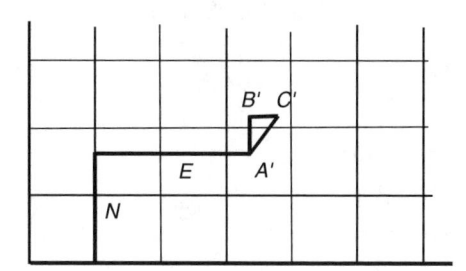

Figure 9.8 The simple cylindrical projection.

Another approach might be to plot the Northings correctly as $R\phi$ and the Eastings as $(R \cos \phi) \lambda$ so that both North–South and East–West have a scale factor of 1. But consider what happens to triangle ABC. If for the sake of argument we set the latitude of $A = 30°$ and $C = 40°$ so that $\delta\phi = 10° = 0.1745$ radians, and we set $\delta\lambda = 15° = 0.2618$ rad and we have a globe of radius $R = 100$ units, then $A'B'$ on the flat $= R\delta\phi = 17.45$ and $B'C' = R\delta\lambda \cos 40 = 20.05$. Tan $(\angle B'A'C') = 20.05/17.45$ or $\angle B'A'C' = 49°$, while the distance $A'C' = 26.58$.

On the globe, however, we have to use the spherical triangle formulae on the triangle ANC in Figure 9.7. In this case, $NC = 50°$, $NA = 60°$, angle $ANC = 15°$ and if we call the angle $AC = b$, then using the cosine formula for spherical triangles, we have

$$\cos b = \cos 50 \cos 60 + \sin 50 \sin 60 \cos 15 = 0.9622 \text{ or}$$

$$b = 15.8° = 0.2758 \text{ radians}$$

Thus, on the sphere, the distance is 27.58; but on the grid it is 26.58. Likewise, using the sine formula the angle *NAC* on the globe is 53.7° rather than 49° as previously derived. Thus, although we have preserved the scale in both the North–South and East–West directions, we have changed the scale along the diagonal and the angle at *A*. We would also have the merdians bending toward the center and not crossing the parallels at right angles. This is inevitable because a curved surface must be distorted in order to represent it on a plane.

The question then is how should δN and δE relate to $R\delta\phi$ and $R \cos \phi \delta\lambda$? There are many possible answers, each one trying to preserve the angles, the distances, or the area or somehow compromise between these three quantities. There are, in general, three approaches, one based on a cylinder wrapped around the sphere, another based on a cone touching or cutting it, and a third based on a plane. Both the cylinder and cone can be cut along a line and opened up into a flat surface (Figure 9.9).

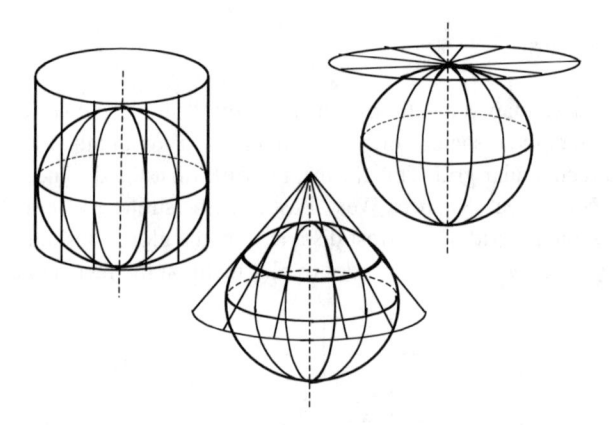

Figure 9.9 Cylinder, cone and plane.

Projections based around these three approaches are called *cylindrical*, *conical*, or *azimuthal* (or *zenithal*). They are normally considered to have the axis of rotation of the Earth as vertical, but they can also be applied obliquely.

9.5 CYLINDRICAL PROJECTIONS

Firstly, let us consider the cylinder that wraps around the equator so that when it is cut along a vertical line and unfolded, it creates a flat surface on which the meridians become parallel lines. The parallels of latitude cut the cylinder in a series of parallel lines that are perpendicular to the meridian lines. Three examples are shown in Figure 9.10, the first preserving distances North–South, the second preserving area, and the third preserving scale around any point.

On the globe, the meridians are convergent, that is, they come together at the poles but always cross the parallels of latitude at right angles. In the ordinary cylindrical projections, the parallels remain parallel and the meridians intersect them at 90°. It is the spacing between the parallels that is adjusted. Consider the small elemental triangle shown as *ABC* in Figure 9.7 with two sides $R \cos \phi \, \delta\lambda$ and $R\delta\phi$ and reproduced in Figure 9.11.

TRANSFORMATIONS

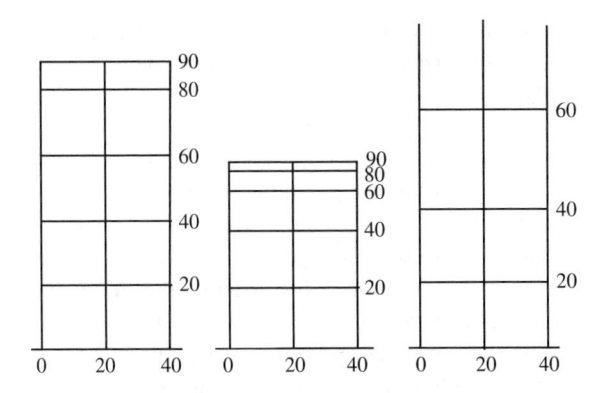

Figure 9.10 Cylindrical equidistant, equal area and conformal.

For the simple cylindrical, the spacing between the parallels is retained and therefore $\delta N = R\delta\phi$. Hence $dN = R\,d\phi$ or $N = \int R d\phi = R\phi + C_n$. For differences in longitude $\delta E = R\delta\lambda$; hence, $dE = R\,d\lambda$ or $E = \int R\,d\lambda = R\lambda + C_e$. C_n and C_e are the constants that must be introduced on integration and ϕ and λ must be measured in radians. Since $N = 0$ when $\phi = 0$ and $E = 0$ when $\lambda = 0$ then $C_n = C_e = 0$. Hence, for the simple cylindrical projection, $N = R\phi$ and $E = R\lambda$.

To preserve the area of the small element, if we retain $E = R\lambda$ when it really ought to be $R \cos\phi\,\lambda$, then we have in effect increased the scale West–East by sec ϕ.

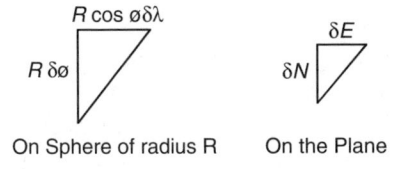

Figure 9.11 Elements on a sphere and plane.

Thus, to preserve area, we must reduce the South–North distances by an amount cos ϕ. In other words, since the area of a triangle is half its base*height, the area of the elemental triangle on the sphere is $(1/2)R^2 \cos\phi\,\delta\phi\,\delta\lambda$. The area on the projection is $(1/2)\,\delta N\,\delta E$. But $\delta E = R\delta\lambda$; hence, $\delta N = R \cos\phi\,\delta\phi$ and

$$N = \int R \cos\phi\,d\phi = R \sin\phi$$

Thus, for the cylindrical equal area projection,

$$E = R\lambda \text{ and } N = R\sin\phi$$

Projections that preserve area are called *equal-area* or *equivalent* or *authalic*. If we want to maintain the shape of the elemental triangle, then since the West–East scale factor is "sec ϕ," then we must do the same for the South–North and thus

$$\delta N = R \sec\phi\,\delta\phi$$

The integral of sec ø was shown in Box 6.5 to be $\log_e(\sec ø + \tan ø)$. This can also be expressed as $\log_e \tan(\pi/4 + ø/2)$. If $dN = R \sec ø\, dø$ then $N = \int R \sec ø\, dø = R \log_e \tan(\pi/4 + ø/2)$.

By maintaining the scale around any point, the shape of small areas is preserved. Projections that maintain this property are known as *orthomorphic* or *conformal*. They have the advantage that angles around any point are preserved so that angles measured in the field with a theodolite can be used directly in the projection. This is particularly important in large-scale topographic mapping.

One consequence is that at the North or South pole, the value for N becomes $\log_e \tan(90°)$ and this is infinitely large. Recalling that this is based on a cylinder wrapped around the equator there is no way that the position of the poles can be plotted on a map when using this form of cylindrical projection.

The cylindrical orthomorphic projection is also known as *Mercator's Projection*. Since angles around any point are preserved and since any straight line drawn on the map cuts all the meridians at the same angle, it means that the bearing along the line is always the same. A line of constant bearing is known as a *rhumb line* or *loxodrome*. Such lines have been particularly important to navigators who can draw a straight line on a chart using the Mercator's Projection and know that by following the bearing of that line they will reach their destination. This line may not be the shortest route; but it may be the easiest to navigate.

The scale at any point on the Mercator's Projection is given by "sec ø". This becomes large at high latitudes; but around the equator it is close to 1. By wrapping a cylinder around a meridian rather than the equator, we have a transverse cylindrical projection.

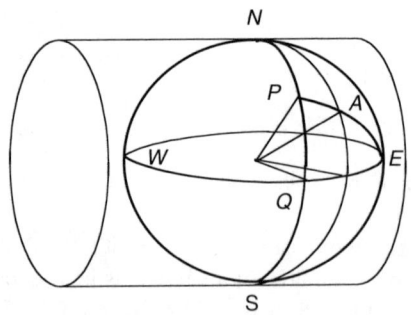

Figure 9.12 The Transverse Mercator.

In Figure 9.12, the cylinder is wrapped around the meridian *NPQS*, known as the central meridian, with the points *E* and *W* being 90° East and West of the central meridian, in effect becoming the poles in the standard case. For any point *A*, *EAPW* is a great circle and the coordinates of *A* can be expressed in terms of *QP* (the equivalent of λ in the normal cylindrical case but here being $ø_p$). The angle *PA* subtended at the center of the sphere is the equivalent of ø in the normal case; but here it needs to be calculated. Let us assume that its value is α.

For the transverse cylindrical equidistant projection (also known as the Cassini projection), the Northings of the point *A* are plotted as $N = Rø_p$, while for the Eastings, $E = R\alpha$. The Cassini projection was much used in early days for topographic mapping. Similarly, if we let $E = R \sin \alpha$, then we have the transverse cylindrical equal area.

If, however, we make $E = R \log_e (\sec \alpha + \tan \alpha) = R \log_e \tan(45° + \alpha/2)$, then we have a conformal projection known as the Transverse Mercator. By extending this principle to an ellipsoid rather than to a sphere and then mapping the globe in strips 3° either side of the central meridians that are chosen every 6° of longitude, a picture of the globe can be built up in what is known as the Universal Transverse Mercator or UTM.

The Transverse Mercator is the most commonly used projection in surveying and is normally related to the spheroid rather than the sphere that gives the best representation of the shape of the earth. As we saw in Chapter 8, the geometry of the ellipse and ellipsoid is more complex than that of the circle and sphere. In essence, there are two radii of curvature that must be considered — ρ and ν. Further discussions of these, and how they affect map transformations are beyond the scope of the present book.

Box 9.1 — Cylindrical Projections

For the Cylindrical Equidistant or Plate Carree

$$E = R\lambda \text{ and } N = R\phi$$

For the Cylindrical Equal Area

$$E = R\lambda \text{ and } N = R\sin\phi$$

For the Cylindrical Orthomorphic or Mercator's

$$E = R\lambda \text{ and } N = R \log_e(\sec \phi + \tan \phi)$$

For the Transverse Cylindricals:
For the Equidistant (The Cassini projection):

$$E = R\alpha \text{ and } N = R\phi_P$$

For the Cylindrical Orthomorphic (The Transverse Mercator):

$$E = R \log_e (\sec \alpha + \tan \alpha) \text{ and } N = R\phi_P$$

9.6 AZIMUTHAL PROJECTIONS

An azimuthal projection is one where the surface onto which points are projected is a plane that is tangential to the sphere. One of the easiest azimuthal projections to imagine is the polar case (known as the zenithal projection) in which either pole (here designated N) becomes the center, with the meridians radiating outward as straight lines. The parallels become circles centered around N, their radius depending on the characteristics that we wish to preserve.

On the sphere, the distance along the surface from the pole to the parallel = $R(\pi/2 - ø) = R\chi$, where ø is the latitude of the point P and χ = colatitude, the angle as measured from the Pole. This distance plots on the flat as the length 'r,' whose value depends on the characteristics of the projection. The angles λ representing the meridians are plotted as their values on the Earth's surface, that is, they are true to scale.

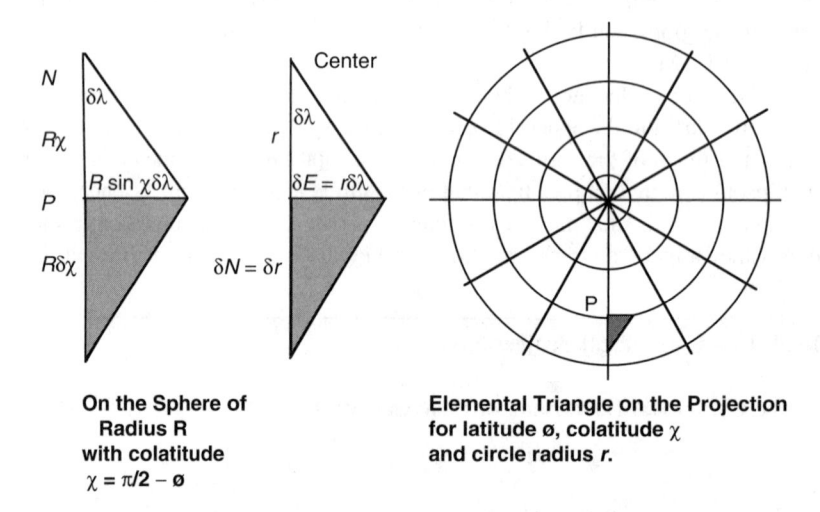

Figure 9.13 Zenithal projections.

On the Sphere of Radius R with colatitude $\chi = \pi/2 - ø$

Elemental Triangle on the Projection for latitude ø, colatitude χ and circle radius r.

Consider a small triangle of width $\delta\lambda$ and height $\delta ø$ on the sphere and on the projection. For the zenithal equidistant, circles are drawn around the Pole with radius $r = R(\pi/2 - ø) = R\chi$. For the equal area projection, we have to make the two elemental triangles in Figure 9.13 of the same area; hence

$$(1/2) R^2 \sin \chi \, \delta\chi \, \delta\lambda = (1/2) r \, \delta r \, \delta\lambda \text{ or } \int R^2 \sin \chi \, d\chi = \int r \, dr$$

This yields $- R^2 \cos \chi + C = (1/2)r^2$ where C is some constant. Since $r = 0$ at the center where $\chi = 0$ and $\cos 0 = 1$, $C = R^2$. Rearranging, $r^2 = 2R^2(1 - \cos \chi)$ and this equals $4R^2\sin^2(\chi/2)$ as shown in Chapter 5. Hence, $r = 2R\sin(\chi/2)$ for the equal area projection.

For the zenithal orthomorphic projection, in order to preserve shape, then from Figure 9.13

$$R \sin \chi \, \delta\lambda/R\delta\chi = r \, \delta\lambda/\delta r \text{ or } (1/\sin \chi) \, d\chi = (1/r) \, dr$$

Integrating, $\log_e(\tan \chi/2) + \text{constant} = \log_e r$. If we call the constant $\log_e C$, then $\text{Log}_e(C \tan \chi/2) = \log_e r$ or $r = C \tan(\chi/2)$. The value of C determines the scale overall and is commonly set so that $C = 2R$. This projection is often known as the *stereographic projection*.

The relationships established above can also be applied to the projection onto a plane that touches the sphere at any point P, not just the pole. The formulae have to be modified as χ and λ used above take on a different meaning. The angle *NPA* in

TRANSFORMATIONS

Figure 9.14 represents the difference in longitude; hence, we must replace λ by Ω (omega), while the angle PA is now what was the colatitude. Thus, we must replace χ by Ψ (psi). The values can be calculated from the sine and cosine formulae for spherical triangles.

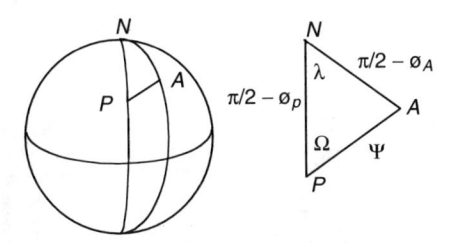

Figure 9.14 Oblique azimuthal.

From the sine formula: $\sin \Omega / \sin (\pi/2 - \phi_A) = \sin \lambda / \sin \Psi$ or, since $\sin (\pi/2 - \phi) = \cos \phi$, $\sin \Omega \sin \Psi = \sin \lambda \cos \phi_A$. From the cosine formula

$$\cos \Psi = \cos (\pi/2 - \phi_A) \cos (\pi/2 - \phi_p) + \sin (\pi/2 - \phi_A) \sin (\pi/2 - \phi_p) \cos \lambda$$

or

$$\cos \Psi = \sin \phi_A \sin \phi_p + \cos \phi_A \cos \phi_p \cos \lambda$$

Hence, we can find Ψ and from that we can find Ω.

The stereographic projection has the characteristic that circles on the surface of the Earth plot as circles on the projection. Hence, when looking for the epicenter of an earthquake, we can plot the time when shockwaves were recorded and this will give an indication as to where the center of the earthquake lay.

Box 9.2 — Zenithal Projections

If the colatitude is $\chi = 90 - \phi$, then

For the zenithal equidistant, $\quad r = R\chi$

For the zenithal equal-area, $\quad r = 2R \sin(\chi/2)$

For the zenithal orthomorphic, $\quad r = 2R \tan(\chi/2)$

9.7 CONICAL PROJECTIONS

The third group of projections is known as the conicals. These are based on the idea of a cone wrapped around the mid-latitudes, it being possible to cut a cone so that its surface can be turned into a flat piece of paper. The cone may touch the sphere — in which case the parallel of latitude where this happens is known as the *standard parallel* — or cut the sphere, in which case there will be *two standard parallels* (Figure 9.15).

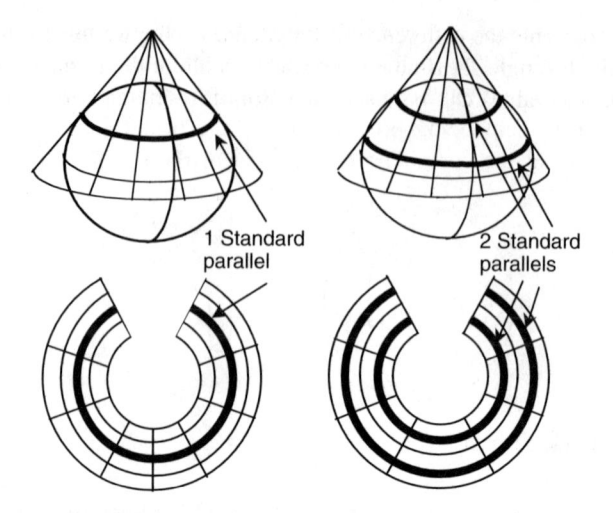

Figure 9.15 Conical projection with 1 or 2 standard parallel.

When the cone is unwrapped and laid flat, each parallel of latitude becomes a part of a set of concentric arcs, while the meridians are lines that radiate from the center. The pole also becomes a circle, whose radius will depend upon the extent to which the cone is pointed.

In fact, the cylindrical and zenithal projections are special cases of the conical, with the angle of the cone being zero for the cylindrical and 180° for the zenithal; if QN in Figure 9.16 is zero, the cone will have become a plane while if QN is infinite in length, then the cone will be a cylinder.

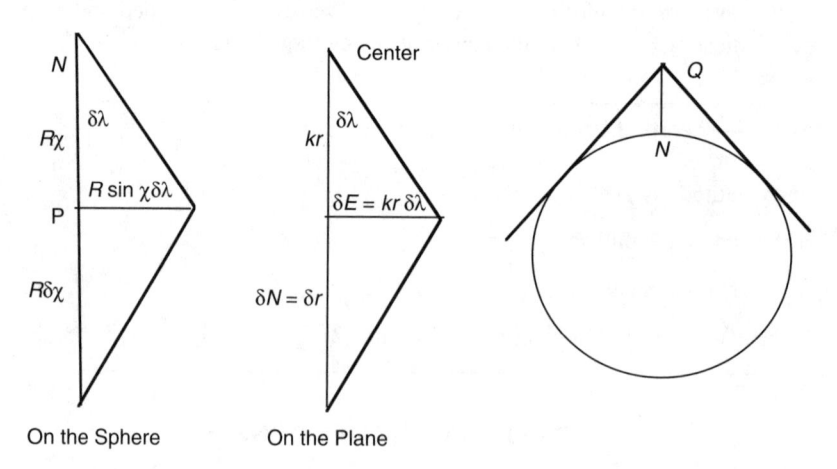

Figure 9.16 Elemental triangles for conical projections.

If we compare the conical case with the zenithal, the radius from the center instead of being r becomes kr where k depends upon the shape of the cone that in turn depends upon the one or two standard parallels.

TRANSFORMATIONS

In the case of any standard parallel (Figure 9.17), the north pole becomes an arc of a circle of radius $C'N'$ (that we will call r_0), while the standard parallel has a correct scale. The meridians become straight lines radiating out from the center, the angles between them being proportional to the true values by 'k' which is called the constant of the cone. The spacing out of the parallels depends on the characteristics of the projection.

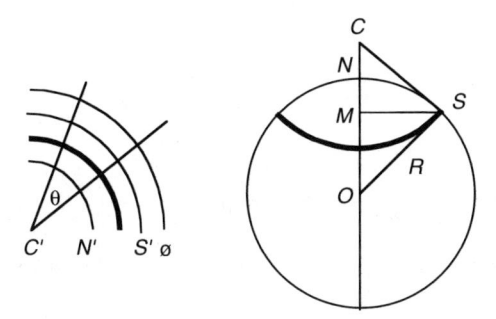

Figure 9.17 Conical equidistant with one standard parallel.

If the separation between any two meridians is λ on the sphere, then on the projection, this angle will have to be reduced by some factor. For the standard parallel with latitude ϕ_s the distance along the parallel is the circumference of a circle of radius SM, which in Figure 9.17 is $R \cos \phi_s$, where R is the radius of the sphere. The whole length of the standard parallel on the sphere is $2\pi R \cos \phi_s$. On the flat, this will be an arc of radius CS, which will equal $R \cot \phi_s$. The angle between the meridians will therefore have to be reduced by a factor 'k' such that

$$k * 2\pi * R \cot \phi_s = 2\pi R \cos \phi_s$$

Thus, 'k' $= \sin \phi_s = \cos \chi_s$ where χ is the colatitude or $(\pi/2 - \phi)$. A difference in longitude of λ becomes $k\lambda$ on the projection (Figure 9.16).

For the conical equidistant with one standard parallel, we need to keep the parallels correctly spaced. Then for any parallel of colatitude χ, the radius of the circular arc that appears on the projection will be CS + the distance along the meridian from the standard parallel $= CS + R(\phi_s - \phi)$. But $CS = R \cot \phi_s$. Thus, if the distance from the center $C' = r$, then

$$r = R \cot \phi_s + R(\phi_s - \phi) = R \tan \chi_s + R(\chi - \chi_s)$$

In particular, for the pole where $\phi = \pi/2$,

$$C'N' = R \cot \phi_s + R(\phi_s - \pi/2)$$

We called this radius r_0; thus $C'N' = r_0 = R \cot \phi_s + R(\phi_s - \pi/2)$. The radius at any other latitude $= r = r_0 + R(\pi/2 - \phi)$.

For the case where there are two standard parallels ϕ_1 and ϕ_2, the length of each parallel on the sphere will be such that

$$kr = R \cos \phi = R \sin \chi$$

where k is the constant of the cone and χ is the colatitude. For the two standard parallels

$$k\{r_0 + R(\chi_1)\} = R \sin \chi_1$$

$$k\{r_0 + R(\chi_2)\} = R \sin \chi_2$$

After suitable manipulation, we obtain

$$r_o = R(\chi_2 \sin \chi_1 - \chi_1 \sin \chi_2)/(\sin \chi_2 - \sin \chi_1), \ r = r_0 + R\chi$$

For the conical equal area projection, the two elemental triangles must be of equal area; hence

$$(1/2)R \ d\chi \ R \sin \chi \delta\lambda = (1/2)kr\delta\lambda\delta r \text{ or } R^2 \sin \chi \ d\chi = kr \ dr$$

Integrating, $C - 2R^2 \cos \chi = kr^2$, where C is the constant of integration and k is not yet determined. Thus $kC - 2kR^2 \cos \chi = k^2r^2$.

Along a standard parallel, distances on the sphere are the same as on the projection. Hence along a standard parallel $kr = R \sin \chi$. If there are two standard parallels (1 and 2), then

$$kC - 2kR^2 \cos \chi_1 = R^2 \sin^2 \chi_1, \ kC - 2kR^2 \cos \chi_2 = R^2 \sin^2 \chi_2$$

Subtracting the two equations, and dividing,

$$2k = (\sin^2 \chi_2 - \sin^2 \chi_1)/(\cos \chi_1 - \cos \chi_2)$$

But $\sin^2 = 1 - \cos^2$;
so $(\sin^2 \chi_2 - \sin^2 \chi_1) = (\cos^2 \chi_1 - \cos^2 \chi_2) = (\cos \chi_1 + \cos \chi_2)(\cos \chi_1 - \cos \chi_2)$.
Or $k = (1/2) (\cos \chi_1 + \cos \chi_2) = (1/2) (\sin \phi_1 + \sin \phi_2)$. where ϕ_1 and ϕ_2 are the latitudes rather than colatitudes of the standard parallels. Substituting for k in the equation $C - 2R^2 \cos \chi = kr^2$,

$$C = R^2(1 + \sin \phi_1 \sin \phi_2)/k = R^2(1 + \cos \chi_1 \sin \chi_2)/k$$

From this, r^2 can be calculated and hence r can be derived. The derivation of the values of C and r for the conical orthomorphic follows the same principles as those for the stereographic or zenithal orthomorphic projection involving natural logarithms and will not be described here. It takes the form

$$r = \{(\sin \chi_1)/k\}\{\tan(\chi/2)/\tan(\chi_1/2)\}^k$$

The results are summarized in Box 9.3.

TRANSFORMATIONS

> **Box 9.3 — Conical Projections**
>
> For the conical equidistant with one standard parallel ϕ_s, the radius of each arc of the projection is r where
>
> $$r = r_0 + R(\pi/2 - \phi) \text{ where } r_0 = R\{\cot \phi_s + (\phi_s - \pi/2)\}$$
>
> The meridians converge at a point and if the angle of longitude between two meridians is λ on the sphere and θ on the projection, $\theta = \sin \phi_s \lambda$. For the conical equidistant, with 2 standard parallels whose co-latitudes are χ_1 and χ_2,
>
> $$r = R(\chi_2 \sin \chi_1 - \chi_1 \sin \chi_2)/(\sin \chi_2 - \sin \chi_1) + R\chi$$
>
> For the conical equal area,
>
> $$k = (1/2)(\cos \chi_1 + \cos \chi_2), \quad C = R^2(1 + \cos \chi_1 \sin \chi_2)/k,$$
> $$r = \sqrt{\{C/k - 2R^2 \cos \chi/k\}}$$
>
> For the conical orthomorphic,
>
> $$r = \{(\sin \chi_1)/k\}\{\tan(\chi/2)/\tan(\chi_1/2)\}^k$$

Further details can be found in standard texts on map projections.

In addition to these basic projections, there is a great variety of alternatives that compromise between the preservation of various angles, small shapes, areas, and distances. The reader should consult other texts for details (see Further Reading).

CHAPTER 10

Basic Statistics

CONTENTS

10.1 Probabilities 153
10.2 Measures of Central Tendency 156
10.3 The Normal Distribution 160
10.4 Levels of Significance 163
10.5 The t-Test..165
10.6 Analysis of Variance 166
10.7 The Chi-squared Test...169
10.8 The Poisson Distribution 170

10.1 PROBABILITIES

So far, we have dealt with matters for which there is, at least in theory, an exact answer — curves and surfaces that have been fitted through points and values such as sin ø have been calculated to as many significant figures as were necessary. In many circumstances, however, things are not exact. Every measure can be subject to a small amount of error, which means that we must look for a "best-fit" solution.

Many measures are of phenomena for which there is no precise answer, such as measures of human preference as in estimates of how people may vote in an election prior to the actual event. We need some way of estimating the reliability of any measurement and some way of handling inconsistencies. In other words, we need statistical measures where a *statistic* is some function of random variables that can be used as an estimator of a population. A *population* (sometimes referred to as a universe) is the complete set of individual components or events from which *samples* are drawn. The sample should be chosen so that its characteristics mimic those of the whole population.

Descriptive statistics are sets of numbers used to summarize a set of known data in a clear and concise way, while *inferential statistics* result from the theory and practice of using statistical data to draw conclusions from random samples. Both processes are *heuristic*, meaning that they are guided by experience and experiment rather than by rigorous logical argument from precisely defined axioms.

The word "*random*" means that the value of the item under consideration cannot be predetermined other than in terms of the probability with which it may occur. In particular, it cannot be determined by what has happened before. Thus, a *random number* is one that cannot be determined from any of the previous numbers that have been selected and therefore it does not follow any particular regular or repetitive pattern. In particular, in a series of random numbers, each number is as likely to occur as often as any other in the series. If this does not happen, then the data are biased.

Processes that can be described by random variables are said to be *stochastic* and are described in terms of *probability*, which is a measure of the degree of confidence that can be had in any event. Thus, when tossing an unbiased coin, there is a 50% probability that it will come down head-side up and 50% that it will be "tails."

If we consider the toss of a coin, we can describe the probability of the outcome as $(ph + qt)$, where 'h' is simply a dummy term meaning "heads" and 't' means "tails"; 'p' is the chances out of 1 that the main event will happen and 'q' is the chances that it will not happen. $q = (1 - p)$ since it is 100% certain that either it will or will not happen. In the case of tossing an unbiased coin, $p = q = (1/2)$.

Thus, after one throw, we obtain $(1/2h + 1/2t) = (h + t)/2$. If we repeat the experiment a second time, we get $\{(h + t)/2\} * \{(h + t)/2\} = \{(h^2 + 2ht + t^2)/4\}$, which tells us that out of 4 possible outcomes (the denominator), there is one chance that we have heads twice (h^2), two chances of getting either a head and a tail or a tail and a head ($2ht$), and one chance of getting two tails (t^2).

If we repeat the process a third time, we obtain

$$\left\{\frac{h+t}{2}\right\}^3 = (h^3 + 3h^2t + 3ht^2 + t^3)/8$$

We can write this as $(h^3t^0 + 3h^2t^1 + 3h^1t^2 + h^0t^3)/8$. It tells us that after three throws, there are 8 possible outcomes, there being one chance in 8 of obtaining three heads (h^3) with no tails (t^0), one chance out of eight of getting three tails (t^3) with no heads (h^0), three chances of getting two heads and one tail ($3*h^2t^1$), and three chances of getting one head and two tails ($3*h^1t^2$).

Note that in the expression $(h^3t^0 + 3h^2t^1 + 3h^1t^2 + h^0t^3)/8$, the sum of the coefficients of the terms $h^m t^n$ is $(1 + 3 + 3 + 1) = 8$, which is the same as the denominator. This confirms that the total probability is 8/8, which is precisely one. In each component $h^m t^n$ the sum of the indices $m + n$ is the same, here, after 3 throws it equals 3.

For four tosses $\left\{\frac{h+t}{2}\right\}^4 = (h^4t^0 + 4h^3t^1 + 6h^2t^2 + 4h^1t^3 + h^0t^4)/16$ and again

$(1 + 4 + 6 + 4 + 1)/16 = 1$. The indices of h and t all add up to 4. There is one chance in 16 of getting four heads, 4 chances of getting three heads, 6 of getting two heads, 4 of getting only one head, and one chance in 16 of getting no heads at all.

The expansion of $\left\{\frac{h+t}{2}\right\}^n$ is known as the *binomial expansion* (Box 10.1) and results in

$$\{h^n t^0 + {}_nC_1 h^{(n-1)} t^1 + {}_nC_2 h^{(n-2)} t^2 + \cdots + {}_nC_{(n-1)} h^1 t^{(n-1)} + h^0 t^n\}/2^n$$

where $_nC_r$ is a short-hand way of writing the number $\dfrac{n!}{(n-r)!r!}$ and $n! = 1*2*3...*(n-1)*n$

Box 10.1 — The Binomial Expansion

$$(p + q)^n = p^n + np^{(n-1)}q + \{n(n-1)/2)\}p^{(n-2)}q^2 + \cdots$$

$$+ \;_nC_r\, p^{(n-r)}q^r + \cdots + npq^{(n-1)} + q^n$$

The sum of the indices of '*h*' and '*t*' or '*p*' and '*q*' = *n*. The sum of the coefficients in the numerator divided by the denominator is 1. If, for example, $n = 100$, we can calculate the chances that on a purely random basis all the 100 throws of the coin will be heads. The chances of this are $1/2^{100}$ — pretty unlikely! The chances of getting 99 heads will be $100/2^{100}$, while for 98 heads they will be $4950/2^{100}$ where $4950 = \dfrac{100*99}{1*2}$.

We can extend the use of the binomial expansion where the probabilities are not 50–50. As an example, all other things being equal, the chances that a Wednesday will be the most rainy day in the week will be 1 in 7. The probability is $(pw + q)$, where '*p*' = 1/7 and '*q*' = 6/7 and '*w*' is a dummy to represent "Wednesday."

For two weeks, we have $(pw + q)*(pw + q)$ or for 3 weeks $(pw + q)^3$. Hence, the chance that a Wednesday is the most rainy day in the week for three weeks running is

$$\left(\dfrac{(w+6)}{7}\right)^3 = \left(\dfrac{w^3}{343} + \dfrac{18w^2}{343} + \dfrac{108w}{343} + \dfrac{216}{343}\right)$$

This means that there is one chance in 343 that Wednesday will be the most rainy day of the week for three weeks, 18/343 chances that it happens twice, 108/343 chances that it happens once, and 216/343 chances that no Wednesdays are the most rainy. (Note that $1 + 18 + 108 + 216 = 343$.)

The coefficients in the expansion of $(p + q)^n$ are given in Box 10.2. For instance, in the second line,

(1 2 1) indicates $1*p^2 + 2*p^1q^1 + 1*q^2$, while in the fifth line, (1 5 10 10 5 1) means

$$(p + q)^5 = p^5 + 5p^4q^1 + 10p^3q^2 + 10\,p^2q^3 + 5p^1q^4 + q^5$$

There is, of course, a pattern in all this, whereby the numbers in the lower row are the sum of the two numbers directly above them — for example, the 8th number in row 15 is $6435 = 3003 + 3432$, the 7th and 8th numbers in row 14.

To obtain the probabilities assuming $p = q = 1/2$, for row 15 we must divide all the numbers by $2^{15} = 32768$ so that the whole probability adds up to 1. Hence, dividing through we obtain 0.00003, 0.00047, 0.0032, 0.0139, 0.0417, 0.0916, 0.1527, 0.1964 with the numbers repeated in reverse order for the remaining 8 coefficients, all adding up to 1.0.

Box 10.2 — The Binomial Coefficients	
$n =$	Coefficients in $(p + q)n$
1	1 1
2	1 2 1
3	1 3 3 1
4	1 4 6 4 1
5	1 5 10 10 5 1
6	1 6 15 20 15 6 1
7	1 7 21 35 35 21 7 1
8	1 8 28 56 70 56 28 8 1
9	1 9 36 84 126 126 84 36 9 1
10	1 10 45 120 210 252 210 120 45 10 1
11	1 11 55 165 330 462 462 330 165 55 11 1
12	1 12 66 220 495 792 924 792 495 220 66 12 1
13	1 13 78 286 715 1287 1716 1716 1287 715 286 78 13 1
14	1 14 91 364 1001 2002 **3003 3432** 3003 2002 1001 364 91 14 1
15	1 15 105 455 1365 3003 5005 **6435** 6435 5005 3003 1365 455 105 15 1

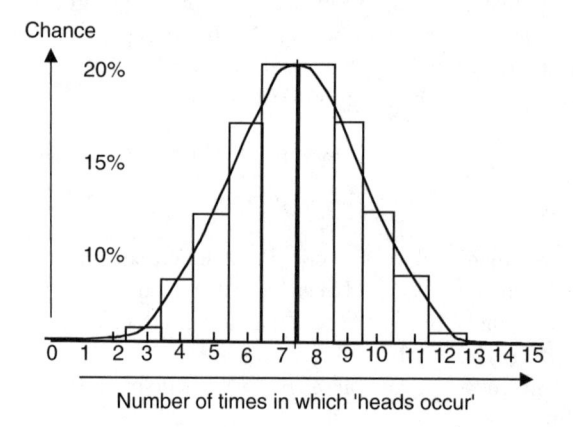

Figure 10.1 A plot of equal probability after 15 events.

10.2 MEASURES OF CENTRAL TENDENCY

In Figure 10.1, the 16 values of the coefficients of p and q have been plotted as a *histogram*, which is a set of contiguous rectangles with width proportional to the size of the class interval and the area proportional to its frequency. They have also been traced in the form of a continuous curve that interpolates the frequency. The chances of there being no "heads" or 15 "heads" in 15 throws are 1 in 32,768 or about 0.00003.

This means that it is probable that if you carried out 32,768 experiments with tossing a coin 15 times, then you would expect, on average, one case where there were 15 consecutive "heads." But it might not happen or it might happen several times. Nothing is certain. If you look at the curve in Figure 10.1, almost the entire area under the curve lies between 3 and 12. The area under the curve is made up of all the separate probabilities and must equal 1 or 100%. We achieved this by dividing the original coefficients by 2^{15}. The areas represented by rectangles in Figure 10.1 are proportional to the overall area and hence graphically represent the probability of the number shown in the bottom line occurring; the total area of these boxes (and under the curve) must be exactly equal to one.

Furthermore, the middle or average value is 7.5. Of course, there cannot be 7.5 "heads" in a single experiment, since we have specified that the outcome is either "heads" or "tails" with a 50% probability of either. But with a series of experiments throwing 15 coins, one would expect that in 7.5 cases there would be at least 8 "heads" and in 7.5 cases there would be 7 or less "heads." In general, if there are 'n' attempts at something for which the probability of it happening is 'p,' then the average number of times that it will happen will be 'np' — 'n' times 'p' (see Box 10.5). Here, $n = 15, p = 1/2$; thus, $np = 7.5$.

In producing a general description of the shape of the curve in Figure 10.1, there is a need to know not only the mean value but how spread out or dispersed the individual answers are likely to be. There are three parameters that are used to express the central point — the *mean*, the *median*, and the *mode*. The median value is the middle number, the mode, the one that most frequently occurs and the mean, which is the average of all the observations.

For example, with two sets of numbers A (1, 5, 24) and B (1, 2, 2, 2, 8, 12, 15, 22), both have a median value of 5 since in the case of B where there is an even number of entities and 5 lies half way between 2 and 8. In B, the mode is 2, while the mean is 8 while for data set A the mean is 10. If the distribution were symmetrical about the middle, then all these values would coincide.

The parameter that is most commonly used to measure how spread out or dispersed is a set of data is known as the *variance*. This is defined as the average sum of the squares of the amount by which each observation differs from the mean. If an observation has a value 'x_i' and the mean value of the population is denoted by the Greek letter for 'm' known as 'mu' (μ), then ($x_i - \mu$) is said to be the *residual* value for x_i.

For 'n' observations, the average value of the sum of the squares of all the residuals is equal to $\Sigma(x_i - \mu)^2/n$ where Σ (the Greek capital letter sigma) means "the sum for all the values of x_i." The variance is usually represented by σ^2, where the symbol σ represents the Greek lower case letter sigma. So $\sigma^2 = \Sigma(x_i - \mu)^2/n$. The square root of the variance, σ is known as the *standard deviation*. Thus, $\sigma = \sqrt{(\Sigma(x_i-\mu)^2/n)}$ for the whole population.

The amount by which any observation differs from the mean when measured in units of standard deviation is called the *Z score* where $z_i = (x_i - \mu)/\sigma$. The mean of a set of z-scores is 0, while the variance and the standard deviation for the set would be 1 (Box 10.3).

In Box 10.4, we give a simple example with two sets of measurements and their means and standard deviations.

The smaller the standard deviation, the more bunched the observations around the mean and therefore the greater confidence there is that the mean represents the

Box 10.3 — Variance and Standard Deviation

For any set of n numbers $x_1, x_2, \cdots, x_i, \cdots, x_n$:
The mean value is

$$(x_1 + x_2 + \cdots + x_i + \cdots + x_n)/n = \Sigma x_i/n = \mu$$

The variance

$$\sigma^2 = \Sigma(x_i - \mu)^2/n$$

The standard deviation

$$\sigma = \sqrt{(\Sigma(x_i - \mu)^2/n)}$$

For each number the Z value

$$z_i = (x_i - \mu)/\sigma.$$

Box 10.4 — Example of Variances

Consider two groups of ten people and their heights in meters.
Group A has heights

1.66, 1.66, 1.67, 1.69, 1.70, 1.71, 1.71, 1.72, 1.73, 1.75

Group B has heights

1.46, 1.52, 1.58, 1.60, 1.68, 1.74, 1.78, 1.84, 1.89, 1.91

The average or mean height in both groups is 1.70 m
The residuals are:
For group A:

−0.04, −0.04, −0.03, −0.01, 0, 0.01, 0.01, 0.02, 0.03, 0.05

For group B:

−0.24, −0.18, −0.12, −0.10, −0.02, 0.04, 0.08, 0.14, 0.19, 0.21

For group A, the variance is 0.00082 and the standard deviation 0.03. For group B, the variance is 0.02226 and the standard deviation ≈ 0.15.

---

best possible estimate of the quantity being measured. The variance and standard deviation are sometimes referred to as *measures of central tendency*.

Box 10.5 demonstrates what common sense should indicate, namely that if each event has a probability of '*p*,' then thereafter '*n*' tries, there are likely to be '*np*'

Box 10.5 — Mean Outcome under the Binomial Distribution

Consider the binomial expansion

$$(q + p)^n = (q^n p^0 + {}_nC_1 q^{(n-1)} p^1 + \cdots + {}_nC_r q^{(n-r)} p^r + \cdots + q^0 p^n)$$

where $(p+q) = 1$, p being the probability that an event will happen, q that it will not happen. $p^0 = q^0 = 1$, while $p^1 = p$ and $q^1 = q$. In the expansion of $(q+p)^n$,

$$_nC_r = \frac{n!}{(n-r)!r!}.$$

Also,

$$_{(n-1)}C_{(r-1)} = \frac{(n-1)!}{\{(n-1)-(r-1)\}!(r-1)!} = \frac{n!}{n} * \frac{r}{(n-r)!r!} = (r/n)\ _nC_r$$

Hence, $_nC_r = (n/r)\ _{(n-1)}C_{(r-1)}$. On average:

The probability of there being no successes is $q^n p^0$.
The probability of one successful outcome is $_nC_1 q^{(n-1)} p^1$.
The probability of "r" successful outcomes is $_nC_r q^{(n-r)} p^r$.
If the total number of expected successful outcomes is A, then

$$A = \{q^n * 0 + {}_nC_1 q^{(n-1)} p * 1 + \cdots + {}_nC_r q^{(n-r)} p^r * r + \cdots + {}_nC_n p^n n\}$$

$$= \{q^n * 0 + {}_nC_1 q^{(n-1)} (np/n) + \cdots + {}_nC_r q^{(n-r)} p^{r-1} * (npr/n) + \cdots + p^{n-1} np\}$$

But $_nC_r = (n/r)\ _{(n-1)}C_{(r-1)}$. Hence,

$$A = np\{q^{(n-1)} + {}_{n-1}C_1 q^{(n-2)} p + \cdots + {}_{n-1}C_r q^{(n-1-r)} p^r + \cdots + p^{(n-1)}\}$$

$$= np(q + p)^{(n-1)} = np \text{ since } (q + p) = 1$$

Thus, on average, there will be $(A=)$ np successful outcomes or the *mean value* $\mu = np$.

successful outcomes. The derivation of the standard deviation is perhaps less obvious and is shown in Box 10.6 to be $\sqrt{(npq)}$ where $q = 1 - p$.

The mean and the variance (and the standard deviation) can be calculated for any set of numbers whatever they represent. Simply add them all together and divide by the number 'n' to obtain the mean. Sum the squares of the differences between each number and the mean, and again divide by 'n' and you have the variance σ^2; take the square root of the variance to obtain the standard deviation σ.

The binomial function allows precise probabilities to be calculated, given certain basic assumptions such as "heads" occurring 1 time out of 2 or rain falling more on a Wednesday 1 time out of 7. The function becomes difficult to handle when the numbers become large and of course it assumes a precise number of outcomes.

Box 10.6 — The Binomial: Variance and Standard Deviation

Given a mean of 'np,' then the original binomial equation shows the probability of the event occurring $np - 1$, $np - 2$, $np - 3$, ...,$np - n$ times. Now the variance σ^2 is the sum of the squares of these differences from the mean multiplied by their relative frequency or

$$\sigma^2 = q^n p^0 (np - 0)^2 + {}_nC_1 q^{n-1} p^1 (np - 1)^2 + \cdots {}_nC_r q^{(n-r)} p^r (np-r)^2$$

$$+ \cdots + p^n (np-n)^2$$

This can be expanded to $L + M + N$ where

$$L = n^2 p^2 (q^n p^0 + {}_nC_1 q^{n-1} p^1 + {}_nC_2 q^{n-2} p^2 + \cdots + {}_nC_r q^{n-r} p^r + \ldots + p^n)$$

$$M = -2 n p (q^n p^0 * 0 + {}_nC_1 q^{n-1} p^1 * 1 + {}_nC_2 q^{n-2} p^2 * 2 + \cdots + {}_nC_r q^{n-r} p^r * r$$

$$+ \cdots + p^n * n)$$

$$N = + (q^n p^0 * 0 + {}_nC_1 q^{n-1} p^1 * 1^2 + {}_nC_2 q^{n-2} p^2 * 2^2 + \cdots + {}_nC_r q^{n-r} p^r * r^2$$

$$+ \cdots + p^n * n^2)$$

The first line, L, gives

$$n^2 p^2 (q^n p^0 + {}_nC_1 q^{n-1} p^1 + {}_nC_2 q^{n-2} p^2 + \cdots + {}_nC_r q^{n-r} p^r + \ldots + p^n)$$

$$= n^2 p^2 (q + p)^n$$

But $(q + p) = 1$, hence the first line $= n^2 p^2$. By expanding and rearranging the terms in the second and third lines M and N we obtain

$M = -2n^2 p^2$ and $N = np + n(n-1)p^2$. Hence,

$$L + M + N = \sigma^2 = n^2 p^2 - 2n^2 p^2 + np + n(n-1)p^2$$

$$= np - np^2 = np(1-p) = npq$$

Hence, $\sigma^2 = npq$. This represents the variance in the probability. The standard deviation $= \sigma = \sqrt{(npq)}$.

10.3 THE NORMAL DISTRIBUTION

The curve that has been drawn in Figure 10.1 passes through discrete points but implies intermediate values. If we accept the possibility of an observation taking any value and still retain the area under the curve to be 1 or 100% probability, then we need a function more complex than the binomial distribution.

BASIC STATISTICS

The fundamental assumption underlying much of statistical analysis is known as the *central limit theorem*, which states that if a sequence of independent identically distributed random variables each has a finite variance then as the number of observed values increases they tend to mirror the probability distribution known as the *Normal Distribution*.

The normal distribution has the form $y = [1/\{\sigma\sqrt{(2\pi)}\}] e^{-\{(x-\mu)^2/2\sigma^2\}}$ or using the z statistics

$$y = \frac{1}{\sigma\sqrt{2\pi}} e^{-(z^2/2)}$$

Box 10.7a — The Normal Curve (1)

Consider Figure 10.2 in which each of the probabilities is represented by a rectangle of width 'w' and height 'y_r'. $y_r = K \,_nC_r \, q^{n-r} p^r$, where '$K$' is some appropriate scaling factor. $_nC_r \, q^{n-r} p^r$ represents the probability of an error of size x_r measured about the mean value.

As shown in the Box 10.6, the mean value $= npw$ and the variance $\sigma^2 = npwqw$. Let x_r be the distance from the mean value to rectangle 'r' $= (r - np)w$. As the number n becomes larger, the widths of the rectangles become narrower, and 'n' becomes infinite and 'w' becomes zero and we finish up with a smooth curve. The probability of an error $x_r + \delta x_r$ is $y_r + \delta y_r = K \,_nC_{r+1} q^{n-r-1} p^{r+1}$. Hence,

$$\delta y_r = K\{ \,_nC_{r+1} q^{n-r-1} p^{r+1} - \,_nC_r q^{n-r} p^r\}$$

$$= K\{q^{n-r-1} p^r\}\{p \,_nC_{r+1} - q \,_nC_r\}$$

But $_nC_{r+1} = n!/\{(n-r-1)!(r+1)!\} = \{(n-r)/(r+1)\} \,_nC_r$. Hence,

$$\delta y_r = K\{ \,_nC_r\}\{q^{n-r} p^r\}\{q^{-1}\}\{(n-r)p - q)/(r+1)\}$$

$$= \{y_r\}\{(np-r)/q(r+1)\}$$

or $\delta y_r/y_r = (np-r)/\{q(r+1)\} = (np-r)w/\{q(r+1)w\}$ based on strips of width 'w.' Now "r" represents the value $(npw + x_r)$ since $(r - np)w = x_r$.
Thus, $(np - r)w = -x_r$ while $(r + 1)w = npw + x_r + w$.
Hence, $\delta y_r/y_r = -x_r/q(npw + x_r + w)$.
Again, multiplying numerator and denominator by w

$$\delta y_r/y_r = -x_r w/\{npwqw + qw(x_r + w)\}$$

$$= -x_r w/\{\sigma^2 + q(x_r + w)w\}$$

As the widths become smaller, we can treat w as δx_r. Hence, $\delta y_r/y_r = -x_r \delta x_r/\{\sigma^2 + q(x_r + \delta x_r) \delta x_r\}$. Thus, $\delta y_r/\delta x_r = -y_r x_r/\{\sigma^2 + q(x_r + \delta x_r) \delta x_r\}$.

Box 10.7b — The Normal Curve (2)

We concluded Box 10.7a with the demonstration that

$$\delta y_r/\delta x_r = -y_r x_r/\{\sigma^2 + q(x_r + \delta x_r)\,\delta x_r\}$$

In the limit, as we increase n and reduce 'w' ($=\delta x$), $q(x_r + \delta x_r)\delta x_r$ tends to zero; hence $dy/dx = -xy/\sigma^2$ and $y = \int(-xy/\sigma^2)\,dx$

This has the solution that $y = k\,e^{-\{(x-\mu)^2/2\sigma^2\}}$, where k is some constant and 'μ' is the mean value. (It is not proven here but may be checked by differentiating y.) The curve is symmetrical about the mean value of $x = \mu$. In order that the total area under the curve is unity and since $\int_{-\infty}^{+\infty} e^{-t^2}\,dt$ can be shown (although not demonstrated in this text) to equal $\sqrt{\pi}$, then $k = 1/\{\sigma\sqrt{(2\pi)}\}$. The equation of the smooth curve, called the normal curve, becomes

$$y = [1/\{\sigma\sqrt{(2\pi)}\}]e^{-\{(x-\mu)^2/2\sigma^2\}}$$

The curve is represented in Figure 10.2. The basis for the derivation of this formula is given in Boxes 10.7a and 10.7b for those who are interested.

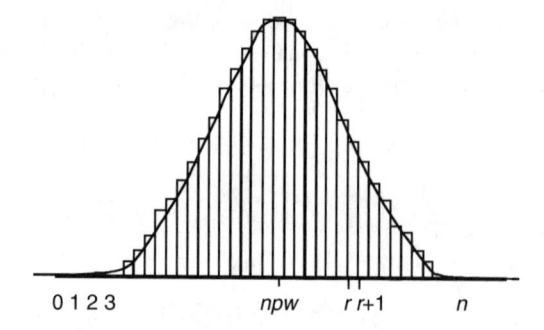

Figure 10.2 A plot of probability for 'n' events.

Those who are prepared to take Box 10.7 as read should note that:

(a) The total area under the curve in Figure 10.2 is 1.
(b) The curve is symmetrical about the value $x = \mu$.
(c) Although the curve goes off to infinity either side of the mean, in practice, over 99% of the observations lie within about 3 standard deviations of the mean.

If we choose to record our measurements about a mean of zero, then $\mu = 0$ and we have

$$y = k\,e^{-\{x^2/2\sigma^2\}} \text{ with } k = 1/\{\sigma\sqrt{(2\pi)}\}$$

This is symmetrical about $x = 0$. The area under the curve is 1 and the variance is σ^2. If we assume that 'x' is a measure of the error in an observation, then the area under the curve between $x = 0$ and r represents the total probability of an error equal to or less than 'r.' This is given by $[1/\{\sigma\sqrt{(2\pi)}\}]\int_0^r e^{-\{x^2/2\sigma^2\}}\,dx$.

BASIC STATISTICS

Unfortunately, there is no easy way to calculate this using elementary techniques. Suffice it to say that using various numerical methods, the sums have been worked out and are available in statistical tables. The ratio between the size of the residual and the standard deviation is illustrated in Table 10.1 where values of r/σ are tabulated against the area under the curve in Figure 10.2.

10.4 LEVELS OF SIGNIFICANCE

Table 10.1 shows that if we measure how far an observation is away from the mean value in units of σ (i.e., r/σ) then, if errors in the measurement are random, the chances are that 0.4987 out of 0.5 (half the area) or 99.7% of the observations will be within the range $-3 < r/\sigma < +3$ or that 99.7% of the observations will fall within 3 standard deviations of the mean. Hence, if we have an observation for which $r/\sigma = 4$, then it is outside the probable range. This suggests that it is almost certain that there is a gross error and the observation does not belong to the group under investigation.

Table 10.1 Partial Areas Under the Normal Curve

r/σ	0.0	0.2	0.4	0.6	0.8
Area	0.0000	0.0793	0.1554	0.2257	0.2881
r/σ	1.0	1.2	1.4	1.6	1.8
Area	0.3413	0.3849	0.4192	0.4452	0.4641
r/σ	2.0	2.2	2.4	2.6	2.8
Area	0.4772	0.4861	0.4918	0.4953	0.4974
r/σ	3.0	3.2	3.4	3.6	3.8
Area	0.4987	0.4993	0.4997	0.4998	0.4999

Before carrying out any statistical analysis, we need to create what is called the *null hypothesis*. A hypothesis is essentially a supposition, an unproved theory, and the null hypothesis is the assumption that there is no significant difference between any particular measure and any other unless we can demonstrate to the contrary. Thus, the object of a statistical test is to prove that something is significantly different from what we would otherwise expect. Statistics cannot prove that anything is true but can indicate the probability that something is significantly different from what might otherwise be anticipated. The key questions then become how to phrase the null hypothesis and how to judge what is significant.

It is almost certain that if $r/\sigma > 3$, there has been an error and the observation can be rejected as highly unlikely to have happened by chance. There is, of course, a very remote possibility that this was a random event since the normal curve extends to infinity. If we have a hypothesis that an event is so rare that it cannot belong to the population under consideration and an observation that really should be included is rejected, then a *Type I Error* is said to occur; if, however, an observation that should be rejected is included then the reverse is the case and a *Type II Error* occurs.

Type I errors can be reduced by increasing the threshold for rejecting observations (known as the *significance level*); but in so doing the chances of making a Type II error are increased. In practice, it is usual to set significance levels at either 95% (the equivalent approximately of $r/\sigma = 2$) or 99% (r/σ is approximately equal to 3).

Consider a data set A that is a sample set of measurements of a particular phenomenon for which there is potentially an infinite possible number of observations. Let the mean and variance of the sample data set A be $\bar{x}$ and S_A^2. In practice, it is highly likely that any small sample will have a slightly different mean and variance from what would be derived from the infinite series.

Let us assume that we have 'n' observations in data set A and let these be $x_1, x_2, \ldots, x_n$ with a mean $\bar{x}$. Then, for the variance of the sample set A, we have

$$S_A^2 = (1/n)\Sigma(x_i - \bar{x})^2 = (1/n) \Sigma (x_i^2 - 2\bar{x}\cdot x_i + \bar{x}^2)$$
$$= (1/n) \Sigma x_i^2 - (2\bar{x}/n) \Sigma x_i + (n/n)\bar{x}^2$$
$$= (1/n) \Sigma x_i^2 - \bar{x}^2$$

If the whole population has a variance σ^2 and mean μ, we would expect that

$$\sigma^2 = (1/n) \Sigma (x_i - \mu)^2 = (1/n) \Sigma x_i^2 - (2\mu/n) \Sigma x_i + \mu^2 = (1/n) \Sigma x_i^2 - 2\mu \bar{x} + \mu^2$$

Subtracting S_A^2 from this, we obtain $(\sigma^2 - S_A^2) = (\bar{x} - \mu)^2$. Thus, the sample variance is smaller than the variance of the infinite set. The means will also differ. The question then arises as to how reliable is the value $\bar{x}$ as an estimate of μ, which is the mean of the whole population.

In fact, if we have 'N' data sets A, B, C, etc., each of size 'n_A,' etc., then they in turn form a set of values with means $\bar{x}_A, \bar{x}_B, \bar{x}_C$, etc., the mean of all of which should be close to the mean of the overall population μ. The estimated variance for the combined N data sets will then be $S^2 = (S_A^2 + S_B^2 + \ldots)/N = \sigma^2/N$ so $S = \sigma/(\sqrt{N})$.

What this shows is that a sample of mean $\bar{x}$ is itself part of a wider distribution arising from many possible samples that could be taken. This larger distribution has the same mean as the total population (μ) and a standard deviation of $\sigma/(\sqrt{n})$, where σ is the standard deviation of the whole population and 'n' is the total size of all the samples that have been taken. In other words, the estimated standard deviation of the mean of a sample A of size 'n_A' will be $(1/\sqrt{n_A})$ times the standard deviation of the sample itself.

We showed above that $(\sigma^2 - S_A^2) = (\bar{x} - \mu)^2$, where σ^2 and μ relate to the whole population and S_A^2 and $\bar{x}$ to the sample. The average value of $(\bar{x} - \mu)^2$ over n samples will be σ^2/n; thus, for any general sample $(\sigma^2 - S^2) = \sigma^2/n$ or $S^2 = (n-1)\,\sigma^2/n$. Thus, if we have a sample with standard deviation S_A, the better estimate for the standard deviation for the population as a whole is $S = S_A\sqrt{(n/(n-1))}$. Box 10.8 summarizes all these relations.

Putting $\sigma^2 = [1/(n-1)]\Sigma (x_i - \bar{x})^2$ (rather than $\sigma^2 = (1/n)\Sigma (x_i - \bar{x})^2$) is called the *unbiased estimate* of the population variance, although it makes little practical difference when the number 'n' is large. The estimate is said to be based on $(n-1)$ *degrees of freedom*, the term referring to the minimum number of parameters necessary to describe completely the state of a system. In statistics, the number of degrees of freedom is equal to the number of independent unrestricted random variables that constitute a particular statistic. Hence, for example, if we have a mean value of ten observations, the mean is fixed, and once 9 of the 10 observations have been chosen, the final tenth observation is already determined. There are thus 9 degrees of freedom.

Box 10.8 — Mean, Variance, and Standard Deviation

For any sample set of n numbers $x_1, x_2, ..., x_i, ..., x_n$ taken from a large population with variance σ^2, standard error σ and mean μ:

The mean value of the sample

$$(x_1 + x_2 + \cdots + x_i + \cdots + x_n)/n = \Sigma x_i/n = \bar{x}$$

The variance of the sample

$$S^2 = \frac{\Sigma(x_i - \bar{x})^2}{n}$$

The standard deviation of the sample

$$S = \sqrt{\frac{\Sigma(x_i - \bar{x})^2}{n}}$$

Also, the standard deviation of the mean $\bar{x}$

$$\frac{\sigma}{\sqrt{n}} = \frac{S}{\sqrt{(n-1)}}$$

10.5 THE *t*-TEST

The normal curve is derived on the basis of the random selection of measures from an infinite population, although in practice it applies equally well as long as the size of the sample is large. When this is not the case, the normal distribution has to be modified but we can still measure in units of standard deviation and calculate $z = (x - \bar{x})/\sigma$. $(x - \bar{x})$ is the difference between an observation and the mean and the division by σ means that we are measuring this difference in units of σ.

An English mathematician William Gosset, writing under the pseudonym of "Student," developed a method to test whether a random sample of normally distributed observations of unknown parameters has the same mean as the population. His statistic is known as '*t*' and his test is known as the *Student's t-test* (Box 10.9).

Box 10.9 — The *t*-test

1. Tests the significance of the difference in means between two data sets.
2. Assumes that the population is normally distributed and that the samples are random.
3. Takes the form $t = (\bar{x} - \mu)\sqrt{n}/S$ and hence measures the difference in means $(\bar{x} - \mu)$ in units of the standard deviation of the mean $\{S/\sqrt{n}\}$.
4. The level of significance or probability can be found from special tables.

The number 't' = $(\bar{x} - \mu)\sqrt{n}/S$, where '$n$' is the number of observations in a sample, '$\bar{x}$' is the sample mean, 'S' is the standard deviation of the sample, and 'μ' is the estimated population mean. Put another way, $t*S/\sqrt{n} = \bar{x} - \mu$.

The test can be used to compare a sample mean with the estimated mean of the whole population. Alternatively, it can be used to test two samples to see if their means differ significantly, in which case '$\bar{x}$' would be the mean of one sample and 'μ' the mean of the other, with 'S' and 'n' being taken either from the combination of the two samples or from one sample, depending on the hypothesis.

The various values of 't' can be found in statistical tables. These show the probability or significance level that arises for various values of 'n' and the calculated value of 't.'.

Table 10.2 Levels of Significance (P) for Values of t (Given 9 Degrees of Freedom)

P	0.9	0.8	0.7	0.6	0.5	0.4	0.3	0.2	0.1	0.05	0.01
t	0.13	0.26	0.40	0.54	0.70	0.88	1.10	1.38	1.83	2.26	3.25

Box 10.10 — Example of t-test

In Box 10.4, we had two groups A and B of ten people whose heights had been measured. Both had mean values of 1.70; but their standard deviations were (A) 0.03 and (B) 0.15.

Does either population differ significantly from an estimated population average of 1.75 m? $\bar{x} = 1.70$, $\mu = 1.75$, $n = 10$, $\sqrt{n} = 3.16$ in both cases.

For group A,
$$t = \frac{(\bar{x} - \mu)\sqrt{n}}{S} = 0.158/0.03 = 5.27$$

For group B,
$$t = \frac{(\bar{x} - \mu)\sqrt{n}}{S} = 0.158/0.15 = 1.05$$

There are 9 degrees of freedom, and (from tables) the level of significance for group A is less than 1%. Hence, one can expect this to happen in less than 1 out of 100 cases. For group B, the result can be expected in more than 0.3 out of 1 or more than 30% of cases, and the result is insignificant. Thus, group A seems to have been a special case, while group B is not abnormal in any particular way.

10.6 ANALYSIS OF VARIANCE

The t-test allows one to compare sample means. The F-test (Box 10.11) may be used to compare the variances between two samples, where

$$F = S_A^2/S_B^2$$

F is the ratio between the variance of sample set A and the variance of sample set B. For very large data sets from the same population, the value of $F \approx 1$.

BASIC STATISTICS

> **Box 10.11 — The F-test**
>
> 1. Is applied to sample variances.
> 2. Assumes that samples A and B are normally distributed.
> 3. Assumes that the variance of A ($= S_A^2$) and B ($= S_B^2$) are independent estimates of the population variance σ^2.
> 4. Takes the form $F = S_A^2/S_B^2$.
> 5. If there are n_A measures in A and n_B in B, then the degrees of freedom are $(n_A - 1)$ and $(n_B - 1)$.
>
> The F-distribution is of particular use in the tests known as the analysis of variance.

For smaller data sets, the significance of the value of F will depend on the number of degrees of freedom in A and B (for instance $(n_A - 1)$ and $(n_B - 1)$). It will also depend on whether one is looking for one of the data sets, such as A, being greater than B or whether one is looking at the probability of their difference being significant regardless of whether A is greater or less than B. If the difference is only relevant in one direction, the test is said to be *one-tailed*; if the absolute difference is important regardless of whether it is greater or less, then the test is said to be *two-tailed*.

In practical terms, the importance lies in where the level of significance is set and it only affects observations that are around the limit of acceptability. Roughly speaking, if 95% of observations are to be within the limit, then for the one-tailed test, the 5% that are outside the limit all lie at one end of the left-hand bell-shaped probability curve in Figure 10.3. For a two-tailed test (the right-hand bell-shaped curve), the two black areas are each 2.5% of the total area below the curve.

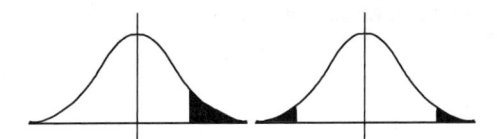

Figure 10.3 One- and two-tailed tests.

The idea of comparing variances leads to a powerful series of tests known as *analysis of variance* or "Anova." If a series of observations are taken, for instance, of the yield of some crop when treated with different types of fertilizers and if there is no significant difference between any of the fertilizers, then one would expect that the variance of each sample would be much the same.

If, however, one fertilizer gave rise to significantly greater yields, then the samples taken from the crop treated with it would consistently have a greater mean. The difference in the mean values could be tested using the *t*-test, while the analysis of variance could help to discriminate between possible factors that are less obvious in a simple test of the mean values. There may be a whole variety of variables, and the analysis of variance can be used to identify which of these is significant.

For example, if the results of a series of 'c' different experiments are recorded for 'r' different samples (see Table 10.3), then they represent "$n = r*c$" different observations for which the overall mean ($\bar{x}$) and standard deviation can be calculated. The first thing to note is that the variance of all the numbers X_{ij} is equal to S^2 where $nS^2 = \Sigma(X_{ij} - \bar{x})^2 = \Sigma\{X_{ij}^2 - 2\bar{x}X_{ij} + (\bar{x})^2\} = \Sigma X_{ij}^2 - 2\bar{x}\Sigma X_{ij} + n(\bar{x})^2$.

But $\bar{x} = (\Sigma X_{ij})/n$. Hence, $n\,S^2 = \Sigma X_{ij}^2 - (\Sigma X_{ij})^2/n$. This is called the *variation*. The point of calculating the variation and variance in this way is that it is then not necessary to calculate each of the individual residuals. More generally,

$$\text{variation} = \Sigma X^2 - (\Sigma X)^2/n.$$

$$\text{variance} = (\Sigma X^2)/n - (\Sigma X/n)^2.$$

Thus, rather than working out all the residuals from the mean and squaring them, we can calculate the sum of each observation squared and subtract the total sum squared divided by 'n' to obtain the variation from which we can obtain the variance and standard deviation.

Referring to Table 10.3, we can calculate the overall variation as $\Sigma(X_{ij}^2) - (\Sigma X_{ij})^2/n$. If we have n observations arranged in n_c columns and n_r rows (so that $n = n_c * n_r$), then if the sum of each column $= T_{cj}$, where $j = 1$ to n_c, then:

The variation between columns is

$$\sum_{j=1}^{j=n_c} \frac{T_{cj}^2}{n_c} - \frac{(\Sigma X_{ij})^2}{n}$$

Table 10.3 Data Classified into Rows and Columns

	Col 1	Col 2	...	Col j	...	Col c	Row total	Row means
Row 1	X_{11}	X_{12}	...	X_{1j}	...	X_{1c}	S_{r1}	$\bar{x}_{r1}$
Row 2	X_{21}	X_{22}	...	X_{2j}	...	X_{2c}	S_{r2}	$\bar{x}_{r2}$
Row 3	X_{31}	X_{32}	...	X_{3j}	...	X_{3c}	S_{r3}	$\bar{x}_{r3}$
⋮	⋮	...	...	...	...	...	...	...
Row i	X_{i1}	X_{i2}	...	X_{ij}	...	X_{ic}	S_{ri}	$\bar{x}_{ri}$
⋮	⋮	...	...	...	...	...	...	...
Row r	X_{r1}	X_{r2}	...	X_{rj}	...	X_{rc}	S_{rr}	$\bar{x}_{rr}$
Column total	T_{c1}	T_{c2}	...	T_{cj}		T_{cc}	ΣX_{ij}	
Column means	$\bar{x}_{c1}$	$\bar{x}_{c2}$	...	$\bar{x}_{cj}$	...	$\bar{x}_{c6}$		$\bar{x}$

BASIC STATISTICS

The variation between rows is

$$\sum_{i=1}^{i=n_r} \frac{T_{ri}^2}{n_r} - \frac{(\Sigma X_{ij})^2}{n}$$

If we add these two variations together, they do not equal the overall variation calculated from $\Sigma(X_{ij}^2) - (\Sigma X_{ij})^2/n$. The difference is known as the *residual variation* or *interaction*. It gives a measure of the independence between the variables.

The data can then be analyzed in terms of whether there is a significant difference between the circumstances that are giving rise to what is shown in rows compared with the whole or whether there are any anomalies in the columns. In each case, the mean and standard deviation can be calculated and if there is no overall significant difference between any of the samples, both the variance between columns and the variance between rows should be about the same as the variance of the whole "n" experiments. This can be verified using the F-test.

10.7 THE CHI-SQUARED TEST

A somewhat similar test is known as the chi-square (Box 10.12) or χ^2 test (from the Greek letter chi or χ). The test considers how often something is observed to happen (f_o) compared with the expected or calculated frequency (f_c), in which $\chi^2 = \Sigma(f_o - f_c)^2/f_c$.

The χ^2 test may be used, for example, if there is a predicted distribution of points on a map (such as cases of cancer near some electrical transmitter) so that the expected numbers can be compared with the observed numbers using the χ^2 test.

Box 10.12 — χ^2 test

1. The test checks whether an observed distribution is probably consistent with estimated values based on a predictive model.
2. The data must be measured at least at a nominal scale or any higher level of measurement. There must be at least two mutually exclusive categories in which the data can be placed, each with a frequency greater than 5.
3. If there are more than 2 categories, then at least 80% of these categories must have at least 5 expected outcomes.
4. If these restrictions cannot be met, then a different test must be used or data categories must be amalgamated so that the criteria are met.
5. The test assumes an observed frequency (f_o) and calculated frequency (f_c).
6. $\chi^2 = \Sigma(f_o - f_c)^2/f_c$.
7. The significance level for the value of χ^2 can be found from statistical tables.

The χ^2 test provides a way in which an observed distribution can be examined to see if it is probably part of a normal distribution — the estimated frequency being provided by the normal function while the observed frequency is what has been measured. The χ^2 test is an example of a *nonparametric test* in that it can be applied to ordinal and nominal data. All the other tests that have been discussed so far have been *parametric* tests, in that they have had parameters such as the mean and standard deviation that can be subject to arithmetic operations.

Box 10.13 — χ^2 Test Example

According to evidence, 15% of properties are sold each year, on average across the whole of a country. In one town with 20,000 properties, only 1000 properties were sold last year. Is that significantly below average?

The expected frequency would be 3000 out of 20,000 or 3000 sold and 17,000 not sold. The observed quantities were 1000 sold and 19,000 not sold (we need to take the non-sales into account to ensure that the total probability (referred to before as $(p + q)$) is equal to 1 or 100%).

$$\chi^2 = (1000 - 3000)^2/3000 + (19{,}000 - 17{,}000)^2/17{,}000$$

$$= 4/3 + 4/17 = 1.57$$

The tables for χ^2 (not shown here) give the chances of this happening as greater than 20%; thus, at least one time in five this may happen. There is nothing particularly significant about the shortfall in house sales.

Parametric tests are based on fairly stringent assumptions, for example, about the nature of the distribution of the population, whereas nonparametric tests, sometimes referred to as "distribution free," are much less demanding. For example, nonparametric tests can be used to compare two lists that have been ranked in order to see if they show similar characteristics, such as, whether secondary or high school level examinations results correlate well with final university degree awards. For further discussion on nonparametric tests, the reader should consult specialist books on statistics.

10.8 THE POISSON DISTRIBUTION

χ^2 can also be used to test other types of distribution, for example, that which bears the name of a French mathematician, Siméon Poisson. Although related to the normal distribution, it is particularly useful when dealing with small samples. It takes the form

$$y = \Sigma \frac{m^r}{r!} e^{-m}$$

Since

$$\frac{m^r}{r!} = \frac{m}{r} * \frac{m^{(r-1)}}{(r-1)!},$$

it means that each term except the first is m/r times the previous term as demonstrated in Box 10.14 and 10.15. 'm' is the mean of a set of samples for which the variance is also 'm' and the standard deviation is $\sqrt{m}$.

BASIC STATISTICS

Box 10.14 — Example of Poisson Distribution (1)

A grid with 10*10 squares was laid over an area, and the number of events (for instance, the number of plants or a particular type of worm) were counted. The results were:

Number of per square	Number of squares	Poisson prediction
0	1	0.5
1	2	2.8
2	6	7.3
3	13	12.7
4	16	16.6
5	19	17.4
6	15	15.2
7	12	11.4
8	10	7.5
9	3	4.3
10	2	2.3
11 or more (=11)	1	2.0
Total	100	

The calculations for the expected frequency according to the Poisson distribution are given in Box 10.15.

Box 10.15 — The Poisson Distribution (2)

Using the data in Box 10.14, which shows the number of squares that have 0, 1, 2, 3, etc., plants in them,

The total number of plants 1*2 + 2*6 + 3*13 + ⋯ + 11*1 = 524.

There are 100 squares; hence

The mean number per square $m = 5.24$. The probabilities are then calculated from

$$y = \sum \frac{m^r}{r!} e^{-m}$$

The value of $(1/e^{5.24}) = e^{-m} = 0.0053$

According to the Poisson formula, the probabilities for $P_0 = 1/(e^{5.24}) = 0.0053$ so out of 100 squares the expected frequency of nil finds is 100*0.0053 = 0.5. Hence

$$P_1 = (m/1)*P_0 = 2.8$$
$$P_2 = (m/2)*P_1 = 7.3$$
$$P_3 = (m/3)*P_2 = 12.7$$
$$P_4 = (m/4)*P_3 = 16.6$$
$$P_5 = (m/5)*P_4 = 17.4$$
$$P_6 = (m/6)*P_5 = 15.2$$

etc.

Thus, if we lay a grid over an area and count the number of events that occur within each square, then we can calculate the probability of their being distributed with an expected frequency. This can be tested to see whether what is predicted by the Poisson model is in conformity with what is found using the χ^2 test.

CHAPTER 11

Best-Fit Solutions

CONTENTS

11.1 Correlation..173
11.2 Regression..176
11.3 Weights..180
11.4 Linearization..182
11.5 Least-Square Solutions..184

11.1 CORRELATION

The tests described so far apply primarily to circumstances in which there is one variable, albeit treated in different ways giving rise to such techniques as the *t*-test and analysis of variance. The assumption is that the data come from the same population.

In practice, and especially in geomatics, at least two things are going on at the same time, for instance, movement in the *x*-direction and independent movement in the *y*-direction. The question then becomes are these somehow interdependent or, in other words, are they correlated? Also, how can we find the best compromise between all the conflicting data?

Given two varibles X and Y where $X = (x_1, x_2, \ldots, x_i, \ldots, x_n)$ and $Y = (y_1, y_2, \ldots, y_i, \ldots, y_n)$ and x_i is possibly related to y_i, then, treating X and Y independently:

Mean of X $\bar{x} = (1/n) \Sigma x$

Mean of Y $\bar{y} = (1/n) \Sigma y$

Variance of X $S^2_x = (1/n) \Sigma (x_i - \bar{x})^2$

Variance of Y $S^2_y = (1/n) \Sigma (y_i - \bar{y})^2$

A measure of the relationship between X and Y is given by $(1/n)\Sigma\{(x_i-\bar{x})(y_i-\bar{y})\}$. This is called the *covariance* of X and Y. The Covariance $(XY) = (1/n)\sum_{i=1}^{i=n}(x_i-\bar{x})(y_i-\bar{y})$ It is often written as "$Cov(XY)$".

Now, we can write $\Sigma(x_i-\bar{x})(y_i-\bar{y})$ as $\Sigma(x_iy_i - x_i\bar{y} - y_i\bar{x} + \bar{x}\bar{y})$ for all $i = 1$ to n. This equals

$$\Sigma(x_iy_i) - (x_1 + x_2 + \cdots + x_n)\bar{y} - (y_1 + y_2 + \cdots + y_n)\bar{x} + n\bar{x}\bar{y}$$
$$= \Sigma(x_iy_i) - n\bar{x}\bar{y} - n\bar{x}\bar{y} + n\bar{x}\bar{y} = \Sigma(x_iy_i) - n\bar{x}\bar{y}$$

Hence

$$Cov(XY) = (1/n)\Sigma(x_iy_i) - \bar{x}\bar{y} = \frac{n\Sigma(x_iy_i) - n\bar{x}\cdot n\bar{y}}{n^2} = \frac{n\Sigma(x_iy_i) - \Sigma x_i \Sigma y_i}{n^2}$$

The covariance has a form similar to the variance, but unlike the latter, which is always positive (being the sum of squares), the covariance may either be positive or negative.

If we divide $Cov(XY)$ by $\sqrt{S_x^2 S_y^2}$ (which is the *geometric mean* of the two variances), we obtain a ratio 'r' that is known as the *coefficient of correlation*. Since

$$S_x^2 = \{\Sigma(x_i-\bar{x})^2\}/n = \{n\Sigma x_i^2 - (\Sigma x_i)^2\}/n^2$$

$$S_y^2 = \{\Sigma(y_i-\bar{y})^2\}/n = \{n\Sigma y_i^2 - (\Sigma y_i^2)\}/n^2,$$

r can be expressed as

$$r = \frac{n\Sigma x_i y_i - \Sigma x_i \Sigma y_i}{\sqrt{\{n\Sigma x_i^2 - (\Sigma x_i)^2\}\{n\Sigma y_i^2 - (\Sigma y_i)^2\}}}$$

Expressing r in this way makes it easier to program the calculation since in tabular form it is relatively easy to sum the raw observations without first having to calculate the means and compute the differences between the means and the observed quantities. This simplifies the way in which the residuals and hence the variance are calculated. (see Table 11.1)

Table 11.1 Framework for Calculating 'r'

Observation	x	y	x^2	y^2	xy
1	*	*	*	*	*
2	*	*	*	*	*
⋮					
n	*	*	*	*	*
Total	Σx	Σy	Σx^2	Σy^2	Σxy

It will be seen from the equation for r that if $x = y$, then the value of 'r' = 1 and if $x = -y$, then 'r' = -1. If we put $2x$ for x or any other multiple of x, then 'r' remains the same, similarly with y. 'r' will have a value of $+1$ for full agreement between x and y, while $r = -1$ for full disagreement; $r = 0$ for no correlation between the two.

BEST-FIT SOLUTIONS 175

Box 11.1 — Example of Correlation

12 observations were made on both x and y:

Obs	x	y	x^2	y^2	xy
1	21	27	441	729	567
2	23	27	529	729	621
3	25	25	625	625	625
4	26	26	676	676	676
5	29	23	841	529	667
6	31	20	961	400	620
7	32	19	1024	361	608
8	35	17	1225	289	595
9	37	14	1369	196	518
10	38	15	1444	225	570
11	41	11	1681	121	451
12	43	9	1849	81	387
Total	$\Sigma x = 381$	$\Sigma y = 233$	$\Sigma x^2 = 12665$	$\Sigma y^2 = 4961$	$\Sigma xy = 6905$

$$r = \frac{n\Sigma x_i y_i - \Sigma x_i \Sigma y_i}{\sqrt{\{n\Sigma x_i^2 - (\Sigma x_i)^2\}\{n\Sigma y_i^2 - (\Sigma y_i)^2\}}}$$

$$= (82860 - 88773)/\sqrt{\{(6819)*(5243)\}}$$

$$= -5913/5979 = -0.989$$

This shows a high level of negative correlation with, since there are two variables X and Y, 12 paired observations; thus, there are $(n - 2) = 10$ degrees of freedom.

The significance level of 'r' can be tested using the appropriate statistical tables. Alternatively, it can be shown by suitable rearrangement of the data that 't' from the "t-test" equates with $r \dfrac{\sqrt{(n-2)}}{\sqrt{(1-r^2)}}$. Thus, by calculating this quantity, the 't' tables can be used.

Box 11.2 — Covariance and Coefficient of Correlation

For two independent variables X and Y:

$$Covariance(XY) = Cov\,(XY) = \frac{n\Sigma(x_i y_i) - \Sigma x_i \Sigma y_i}{n^2}$$

$$Coefficient\ of\ Correlation\ \ r = Cov(XY)/\sqrt{\{S^2_x S^2_y\}}$$

$$r = \frac{n\Sigma x_i y_i - \Sigma x_i \Sigma y_i}{\sqrt{\{n\Sigma x_i^2 - (\Sigma x_i)^2\}\{n\Sigma y_i^2 - (\Sigma y_i)^2\}}}$$

$$-1 \leq r \leq +1$$

11.2 REGRESSION

For certain data sets, it is probably easier to test levels of correlation by viewing them pictorially. Consider a set of points for which there are values of X and Y in pairs with (x_A, y_A), and (x_B, y_B), etc., as shown in Figure 11.1. They appear to fall approximately along a straight line. There appears to be some degree of correlation between them and indeed some linear function, which gives a good approximation to the relationship between the set of points X and their partners Y.

The relationship could be expressed by calculating the coefficient of correlation between them (r). Expressing the relationship in the form of $y = mx + c$ provides additional information. The process of choosing the best-fit relationship between Y and X is called *regression*, in this case *simple linear regression*.

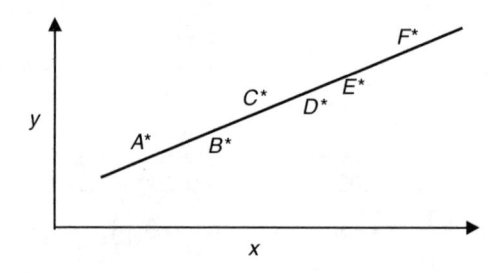

Figure 11.1 Regression line.

The line that has been drawn in Figure 11.1 does not actually pass through any of the points. It could have done, but then it might not. Let us exaggerate the situation as in Figure 11.2.

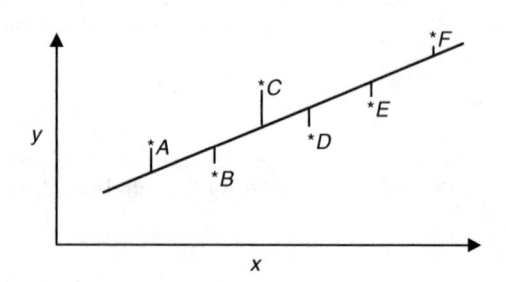

Figure 11.2 Residuals from the regression line.

Each point for which there has been a pair of observations (x_A, y_A), (x_B, y_B), etc., there will be observed $(_o)$ quantities $(_ox_i, _oy_i)$, where 'i' is any point. There will also be a set of computed $(_c)$ values such as $(_cx_i, _cy_i)$ that are derived from the equation for the straight line.

If we assume that the Y values are dependent on the X values and that $_ox_i = {_c}x_i$, implying that the X values are correctly determined, then $_oy_i$ will approximately

BEST-FIT SOLUTIONS

equal $_Cy_i$. In practice, there may be small differences as shown by the short vertical lines above or below the points A, B, C, etc., in Figure 11.2. If the line has the form

$$y = mx + c$$

then

$$_Oy_i = m*_Ox_i + c + \varepsilon_i$$

where 'ε' or the Greek lower case letter epsilon represents a small amount and is sometimes called the *error term* $\varepsilon_i = {_Oy_i} - {_Cy_i} = {_Oy_i} - c - m*_Ox_i$.

As can be seen from Figure 11.2, ε will sometimes be positive and sometimes negative and in general will be small if there is a good correlation between X and Y. The value of ε' will depend not only upon the various values of X and Y but also upon the choice of the two constants 'm' and 'c' for the regression line.

If we assume that 'ε' is normally distributed, the most likely values for 'm' and 'c' will be those that ensure that $\Sigma\varepsilon^2$ is a minimum. This is known as the *least-square solution*. This fundamental assumption occurs in all kinds of problems in measurement science. As can be seen from the formula for the coefficient 'r', the maximum correlation is achieved when the denominator in the expression for 'r' (i.e., the value of $\sqrt{S_x^2 S_y^2}$) is as small as possible. This occurs when the variance is a minimum.

In the general case, we need to choose 'm' and 'c' so that $\Sigma\varepsilon^2$ and hence $\Sigma(y-c-mx)^2$ is a minimum. This means that

$$\Sigma(y^2 + c^2 + m^2x^2 - 2cy - 2mxy + 2cmx)$$

is a minimum or

$$\Sigma y^2 + nc^2 + m^2\Sigma x^2 - 2c\Sigma y - 2m\Sigma xy + 2cm\Sigma x$$

is a minimum. As 'm' and 'c' vary, the minimum will occur when the partial differentials with respect to 'm' and 'c' are zero. This will happen when:

(Differentiating with respect to 'c') $2nc - 2\Sigma y + 2m\Sigma x = 0$

(Differentiating with respect to 'm') $2m\Sigma x^2 - 2\Sigma xy + 2c\Sigma x = 0$

Hence

$$m\Sigma x + cn = \Sigma y$$

and

$$m\Sigma x^2 + c\Sigma x = \Sigma xy$$

Thus, we can calculate 'm' and 'c' as in Box 11.3.

It may be that Y and X are not linear functions of each other. For example, Y may increase in the form of y^k where k is some constant (e.g., 2 or 3 so that X is related

Box 11.3 — Linear Regression

For two variables X and Y where Y is dependent on X, the best-fit straight line takes the form $y = mx + c$, where

$$m = \frac{n\Sigma(x_i y_i) - (\Sigma x_i)(\Sigma y_i)}{n\Sigma x_i^2 - (\Sigma x_i)^2}$$

$$c = \frac{(\Sigma y_i)(\Sigma x_i^2) - (\Sigma x_i)(\Sigma x_i y_i)}{n\Sigma x_i^2 - (\Sigma x_i)^2}$$

For example, using the data in Box 11.1

$$\Sigma x = 381, \quad \Sigma y = 233$$

$$\Sigma x^2 = 12665, \quad \Sigma y^2 = 4961, \quad \Sigma xy = 6905$$

$$m = (82860 - 88773)/(151980 - 145161)$$

$$= -5913 / 6819 = -0.87$$

$$c = (2950945 - 2630805)/(151980 - 145161)$$

$$= 320140/6819 = 46.95$$

Thus, $y = -0.87x + 46.45$

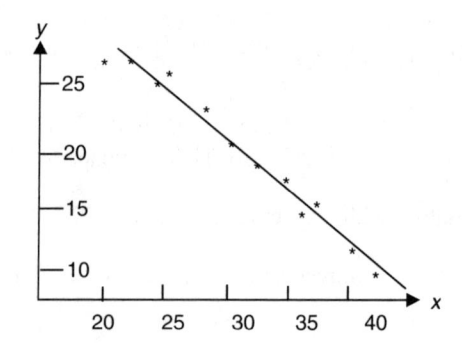

Figure 11.3 Example of regression line.

to Y^2 or Y^3). We discuss aspects of linearization in Section 11.4 but, for the present, we note a relatively simple solution.

If we consider F rather than Y as the function that is dependent on X and if we let $F = \log(y^k) = k \log y$, then we can still plot the regression line between X and F by using $(\log y_i)$ instead of (y_i), the value k merely affecting the slope of the line.

BEST-FIT SOLUTIONS

Plotting one set of values on a logarithmic scale does not invalidate the process as long as the relationship is correctly interpreted.

The principles enunciated above also apply in three dimensions where X and Y are freely independent variables and Z is dependent on X and Y. If Z is some function of X and Y in which $Z = f(x, y)$, then we have a surface. In particular, if the relationship is linear, then the surface is a plane. In general, the surface representing Z is known as the *trend surface*. Trend surface analysis is an extension of linear regression to three or more dimensions.

The technique of finding the best fit by minimizing the variance of the residual values can be extended to the combination of different measurements subject to different conditions. This is possible because as shown in Boxes 11.4 and 11.5, if

$$X = ax + by + cz + \cdots$$

where a, b, c are any numbers, and if

the means of x, y, z are, respectively, $\bar{x}, \bar{y}, \bar{z}$, etc., and

the corresponding standard deviations are S_x, S_y, S_z, etc.,

then

Mean of X is $\bar{X} = a\bar{x} + b\bar{y} + c\bar{z} + \cdots$

Variance of X is $a^2 S^2_x + b^2 S^2_y + c^2 S^2_z + \cdots$

Box 11.4 — Linked Independent Variables — Means

Consider 'm' observations of x and 'n' of y that are combined to form X such that $X = ax + by$. There are '$m*n$' possible values for X since there are 'm' ways of choosing 'x' and 'n' ways of choosing 'y'. Let $x_i = \bar{x} + r_i$ and $y_i = \bar{y} + \rho_i$. Thus $\Sigma r_i = 0$ and $\Sigma \rho_i = 0$ (because $(1/m)\Sigma x_i = \bar{x}$ and $(1/n)\Sigma y_i = \bar{y}$). Also $(1/m) \Sigma r_i^2 = S^2_x$ and $(1/n)\Sigma \rho_i^2 = S^2_y$ (since r_i and ρ_i are the residuals of x and y about their individual means). Then

$$\bar{X} = \frac{1}{mn} \sum_{i=1}^{m} \sum_{j=1}^{n} (ax_i + by_j)$$

$$= \frac{1}{mn} \sum_{i=1}^{m} \sum_{j=1}^{n} (a\bar{x} + b\bar{y} + ar_i + b\rho_j)$$

$$= \frac{1}{mn} \sum_{i=1}^{m} \sum_{j=1}^{n} (a\bar{x} + b\bar{y}) \text{ since both } \Sigma r_i \text{ and } \Sigma \rho_i = 0$$

$$= a\bar{x} + b\bar{y}$$

What is true for x and y must also be true for x and y and z, etc. When summing independent variables, the means are additive.

> **Box 11.5 — Linked Independent Variables - Variance**
>
> Following on from Box 11.4:
> The variance of $X = S^2$ where:
>
> $$S^2 = \frac{1}{mn} \sum_{i=1}^{m} \sum_{j=1}^{n} (ax_i + by_j - a\bar{x} - b\bar{y})^2 = \frac{1}{mn} \sum_{i=1}^{m} \sum_{j=1}^{n} (ar_i + b\rho_j)^2$$
>
> With a little manipulation, this gives us $S^2 = a^2 S^2_x + b^2 S^2_y$. Once again, what is true for x and y will also be true for x and y, z, etc. When summing independent variables, the variances are additive. Thus, if $X = ax + by + cz + \cdots$, where a, b, c are numbers and x, y, and z are related variables and subject to all possible combinations, then
>
> $$\text{Mean of } X \text{ is } \bar{X} = a\bar{x} + b\bar{y} + c\bar{z} + \cdots$$
>
> $$\text{Variance of } X \text{ is } a^2 S^2_x + b^2 S^2_y + c^2 S^2_z + \cdots$$

In the derivations given in Boxes 11.4 and 11.5, if $a = b = c$, etc. $= 1$, then for all the possible combinations of x, y, and z, etc.

$$\text{Mean of } X \text{ is } \bar{x} + \bar{y} + \bar{z} \cdots$$

$$\text{Variance is } S^2_x + S^2_y + S^2_z + \cdots$$

11.3 WEIGHTS

In what we have explored above, the assumption has been that all the observations are equally reliable and therefore the errors in each quantity have been distributed normally with the same value for their standard deviations σ. Sometimes, the observations are of different reliability and therefore we may wish to give them different *weight*, in effect leaning more heavily toward one set of observations than another.

Thus, if we give twice as much weight to y than to x and three times as much to z, then the weighted mean will be $(x + 2y + 3z)/6$. With weights 'w_1,' 'w_2,' and 'w_3,' the weighted mean is given by

$$\frac{w_1 x + w_2 y + w_3 z}{w_1 + w_2 + w_3}$$

Weights are relative values, not absolute values. From a theoretical perspective, according to the normal distribution, the probability of an error $(x_i - \bar{x})$ is proportional to $e^{-(x_i - \bar{x})^2 / 2\sigma^2}$. Thus, the weights of numbers should be inversely proportional to the variances of the observations, that is,

$$\text{the weight of observation '}i\text{'} = w_i = K/\sigma_i^2$$

where K is the same number for all observations. If we put $K = \sigma^2 = \Sigma \sigma_i^2$, which is the variance of the whole population, then the probability of an error $(x_i - \bar{x})$ will be

BEST-FIT SOLUTIONS

$e^{-w_i(x_i - \bar{x})^2/2\sigma^2}$ and for the whole set of n values $x_1, x_2, \ldots, x_i, \ldots, x_n$ the probability will be $e^{-\Sigma w_i(x_i - \bar{x})^2/2\sigma^2}$.

This quantity will be a maximum (to give the maximum probability) when we choose 'm' so that $\Sigma w_i(x_i - m)^2$ is a minimum. Differentiating with respect to m, we obtain $2\Sigma w_i(x_i - m)$ and this must equal zero for a minimum value. Hence, $2\Sigma w_i x_i - 2\Sigma w_i m = 0$ or $m = \Sigma w_i x_i / \Sigma w_i$. Thus by using μ, which is the weighted mean of $x = \Sigma w_i x_i / \Sigma w_i$ we have the most probable value for observations of known weight. As we might have expected, the weighted mean is the most probable value of a set of observations. Note that if all observations are of the same weight, $\Sigma w_i = n$ and the mean is

$$\Sigma x_i / n$$

For weighted observations, the variance becomes

$$\sigma^2 = \frac{1}{n-1} \sum_{i=1}^{i=n} w_i(x_i - \mu)^2$$

Box 11.6 — Example of a Weighted Mean

A distance is measured by three different methods and the results are estimated as:

    a 19.412 ± 0.005, b 19.417 ± 0.008, c 19.419 ± 0.010

The estimated standard deviation for observation 'a' is 0.005 and the variance is 0.000025, and its inverse is 40,000 = w_a. For 'b' the inverse w_b = 15,625 and for 'c' w_c = 10,000.

Allocating weights according to these inverse variances and reducing the calculations by starting at 19.410, so that

$$a = 19.410 + 0.002, \quad b = 19.410 + 0.007, \quad c = 19.410 + 0.009.$$

The weighted mean is

$$19.410 + \frac{(0.002 * w_a + 0.007 * w_b + 0.009 * w_c)}{(w_a + w_b + w_c)}$$

$$= 19.410 + (80 + 109.375 + 90)/(65625)$$

$$= 19.410 + 0.004 = 19.414$$

The new residuals are 0.002, 0.003, and 0.005, while $(n-1) = 2$. Hence

$$\sigma^2 = \frac{1}{n-1} \sum_{i=1}^{i=n} w_i(x_i - \mu)^2$$

$$= (1/2) \frac{(0.002^2 * w_a + 0.003^2 * w_b + 0.005^2 * w_c)}{(w_a + w_b + w_c)}$$

giving σ = 0.002. Thus, a better estimate is that the distance = 19.414 ± 0.002.

11.4 LINEARIZATION

Earlier we pointed out that in regression, we can transform the data and still obtain a valid measure of correlation. The transformation from ordinary numbers into their logarithmic equivalent is one example and is used later in the adjustment of a geometric figure referred to as a braced quadrilateral. The process of converting complex relationships into linear combinations of variables is sometimes referred to as linearization.

Sir Isaac Newton, for example, developed a method to solve quite complex non-linear equations in which we reduce the problem to one dimension and repeat the process iteratively until we have a satisfactory solution. In Box 2.9, we showed how to obtain a square root by iteration, although we did not explain why it worked.

Consider the general case of $f(x) = k$ where 'k' is a constant and $f(x)$ is a function of x such as the polynomial $ax^3 + bx^2 + cx$. Solving the equations $ax^3 + bx^2 + cx = k$ can be complicated. In fact, Newton's method for solving equations is particularly useful where the polynomial is of fifth or higher degree since there are no general algebraic methods for solving higher-order equations.

However, if we can take a guess at an approximate solution (in the case of a cubic, there will of course be three possible answers, although some may involve imaginary numbers), then we can calculate

$$x_{new} = x_{old} - \{f(x_{old}) - k\}/f'(x_{old})$$

Box 11.7 — Newton's Method for Solving Polynomials

In Chapter 6 in Table 6.1, we gave some values for the function $y = 1 + 9x - 6x^2 + x^3$; $dy/dx = 9 - 12x + 3x^2$. What is the value of x for which $y = 0$? From Table 6.1, an answer must lie between $x = -0.5$ and 0. Let us try $x = -0.4$ and use $x_{new} = x_{old} - \{f(x_{old}) - k\}/f'(x_{old})$ where $k = 0$ and $y = f(x)$. Our next attempt should be

$$x_1 = -0.4 - \{1 - 3.6 - 0.96 - 0.064\}/\{9 + 4.8 + 0.48\}$$

$$= -0.4 + 3.624/14.28 = -0.146. \text{ So now try } x = -0.146$$

$$x_2 = -0.146 - \{1 - 1.314 - 0.128 - 0.003\}/\{9 + 1.752 + 0.064\}$$

$$= -0.146 + 0.445/10.816 = -0.105. \text{ So now try } x = -0.105$$

$$x_3 = -0.105 - (1 - 0.945 - 0.066 - 0.001\}/\{9 + 1.260 + 0.033\}$$

$$= -0.105 + 0.012/10.293 = -0.104$$

To three decimal places, the solution is -0.104.

BEST-FIT SOLUTIONS

We then repeat this process over and over until we have a sufficient number of significant figures for the answer to be acceptable. In fact, this is what we did in Box 2.9 where we had $f(x) = x^2 = k$ where k is 27392834. For this, $f'(x) = 2x$ and we set $x_{new} = x_{old} - \{(x_{old})^2 - k\}/2x_{old} = (1/2)\{x_{old} + k/x_{old}\}$.

In Chapter 6, we introduced Maclaurin's Theorem in which

$$y = f(0) + x^1 f'(0)/1! + x^2 f''(0)/2! + x^3 f'''(0)/3! + \cdots + x^n f^n(0)/n!$$

Another 18th-century mathematician, Brook Taylor, has given his name to an extension of this theorem, which states that if $f(x)$ is a polynomial of the form $a_0 + a_1 x + a_2 x^2 + \cdots + a_n x^n$; then, if δ is any number

$$f(x + \delta) = f(\delta) + x f'(\delta)/1! + x^2 f''(\delta)/2! + x^3 f'''(\delta)/3! \ldots$$
$$+ x^n f^n(\delta)/n!$$

where f^n is the nth derivative of $f(x)$. By putting $\delta = 0$ we have the same expression as in the Maclaurin Theorem.

To indicate the proof of Taylor's Theorem, given that

$$f(x) = a_0 + a_1 x + a_2 x^2 + \cdots + a_n x^n$$

we have

$$f'(x) = a_1 + 2a_2 x + \cdots + na_n x^{n-1}$$
$$f''(x) = 2a_2 + 3*2a_3 x + \cdots + n*(n-1)a_n x^{n-2}, \text{ etc.}$$

Given $f(x + \delta) = a_0 + a_1(x + \delta) + a_2(x + \delta)^2 + \cdots + a_n(x + \delta)^n$,
Then $f'(x + \delta) = a_1 + 2a_2(x + \delta) + \cdots + na_n(x + \delta)^{n-1}$, etc.,
$$f''(x + \delta) = 2a_2 + 6a_3(x + \delta) + \cdots + n*(n-1)a_n(x + \delta)^{n-2},$$

etc., for all values of x. So, putting $x = 0$,

$$f(\delta) = a_0 + a_1(\delta) + a_2(\delta)^2 + \cdots + a_n(\delta)^n$$
$$f'(\delta) = a_1 + 2a_2(\delta) + \cdots + na_n(\delta)^{n-1}$$
$$f''(\delta) = 2a_2 + 6a_3(\delta) + \cdots + n*(n-1)a_n(\delta)^{n-2}, \text{ etc.}$$

Combining all these, we can obtain Taylor's Theorem, which states that

$$f(x + \delta) = f(\delta) + x f'(\delta)/1! + x^2 f''(\delta)/2! + x^3 f'''(\delta)/3! \ldots + x^n f^n(\delta)/n!$$

Then, we can also express this as

$$f(x + \delta) = f(x) + \delta f'(x)/1! + \delta^2 f''(x)/2! + \delta^3 f'''(x)/3! \ldots + \delta^n f^n(x)/n!$$

If δ is small, then ignoring terms in δ^2 and rearranging the above,
$\delta = \{f(x + \delta) - f(x)\}/f'(x)$ leading to

$x_{new} = (x + \delta) = x_{old} - \{f(x_{old}) - k\}/f'(x_{old})$, where 'k' is the value of $f(x)$ that we are seeking.

In fact, Taylor's expansion can be extended for functions other than polynomials, including functions of the form $z = f(x,y)$. This involves partial derivatives of the form $\dfrac{\partial^{(r+s)}}{\partial x^r \partial y^s} (f(x, y))$, which means all the combinations of the partial derivatives up to $n = r+s$. As we saw in Chapter 6, a partial derivative is the differential of an expression in terms of one variable, all other variables being treated as constants. We indicate partial differentiation by using the curly delta symbol ∂. We do not intend to explore higher-order partial derivatives because, fortunately, we can stick with only the first derivatives if we have made a reasonable approximation; if not, we shall just have to repeat the process several times. This gives us

$$f\{x + \delta x, y + \delta y\} = f(x,y) + (\partial f/\partial x)\delta x + (\partial f/\partial y)\delta y + \text{some small terms}$$

where $\partial f/\partial x$ means the differential of $f(x,y)$ with respect to x, with y being treated as a constant.

We shall not demonstrate the proof of this here, but simply note that if we can ignore the terms of higher order, then we are left with a linear relationship of the form $f_{new} = f_{old} + (\partial f_{old}/\partial x)\, \delta x + (\partial f_{old}/\partial y)\, \delta y$, where f_{old} is the value that we use when inserting the initial or iterated values into the function $f(x, y)$ and its partial derivative with respect to x and the partial derivative with respect to y. We shall work through an example of how all this applies when we discuss least-square adjustment in 11.5.

11.5 LEAST-SQUARE SOLUTIONS

Returning to the ideas behind regression we have seen that, given a set of observations in which each observation or measurement may contain a small amount of error, we can obtain the most likely value of a quantity by distributing the residuals on the basis of minimum variance or the "least squares." Application of this principle assumes that the errors are normally distributed and of equal weight or that appropriate weights can be identified.

The weighted mean gives the overall most likely value and the probable errors are of the form $e^{-\{w_i(x_i - \bar{x})^2/2\sigma^2\}}$. This will be a maximum when $\Sigma w_i(x_i - \bar{x})^2$ is a minimum. We can extend this idea not just to fitting lines or surfaces, but when dealing with a set of observations various combinations of which are subject to a set of conditions that can be determined. In the general case, if we have 'n' observed quantities $x_1, x_2, \ldots, x_i, \ldots, x_n$ and 'm' independent relationships of the form $ax_1 + bx_2 + cx_3 + \cdots + nx_n = -L$, where L is a numerical value, we can express these in the form shown in Table 11.2.

Table 11.2   Conditions to be Satisfied

1:	$a_1x_1 + b_1x_2 + c_1x_3 + \cdots + f_1x_i + \cdots + n_1x_n + l_1 = 0$
2:	$a_2x_1 + b_2x_2 + c_2x_3 + \cdots + f_2x_i + \cdots + n_2x_n + l_2 = 0$
$\vdots$	
i:	$a_ix_1 + b_ix_2 + c_ix_3 + \cdots + f_ix_i + \cdots + n_ix_n + l_i = 0$
$\vdots$	
m:	$a_mx_1 + b_mx_2 + c_mx_3 + \cdots + f_mx_i + \cdots + n_mx_n + l_m = 0$

BEST-FIT SOLUTIONS

Table 11.2 in matrix form can be expressed as **MX + L = 0**, where **M** is the (m∗n) matrix composed of the a's, b's, and c's, etc., and **L** is the vector of constants. In Geomatics, the observed quantities are normally angles, distances, or time; but the technique arises in all kinds of applications from optimizing financial balance sheets through to quality control procedures. The key point is that linear relationships can be established between the independent variables; where this is not immediately so, the relationships will need to be "linearized," an example of which is discussed later.

Assuming that the observations are of equal weight, and that there are small errors in each of the observations x_i, then for equation 'j' in Table 11.2, let $v_j = a_j x_1 + b_j x_2 + c_j x_3 + \cdots + f_j x_i + \cdots + n_j x_n + l_j$, where j takes the values 1, 2, $\cdots$, m. If we assume that we want to minimize the errors in each of the n equations, then we have to make the sum of the squares of the quantities 'v_j' as small as possible by suitably amending each of the values of 'x_i.' This will give us the most likely consistent solution.

'v_j' has (n + 1) terms so that 'v_j^2' would have $(n + 1)^2$ terms so that the sum of all these, Σv_j^2, would have $m*(n + 1)^2$ terms. However, we can simplify matters. To minimize Σv_j^2 for x_1, we need to make the partial derivative $\partial(\Sigma v_j^2)/\partial x_1 = 0$. To minimize Σv_j^2 for x_2, we need to make $\partial(\Sigma v_j^2)/\partial x_2 = 0$, etc. Now

$$\frac{\partial(v_1^2 + v_2^2 + v_3^2 + \cdots)}{\partial x_1} = 2\left\{ v_1 \frac{\partial v_1}{\partial x_1} + v_2 \frac{\partial v_2}{\partial x_1} + \cdots + v_m \frac{\partial v_m}{\partial x_1} \right\}$$

$$= 0 \quad \text{when things reach a minimum}$$

$$v_1 \frac{\partial v_1}{\partial x_1} = a_1\{a_1 x_1 + b_1 x_2 + c_1 x_3 + \cdots + f_1 x_i + \cdots + n_1 x_n + l_1\}$$

$$v_2 \frac{\partial v_2}{\partial x_1} = a_2\{a_2 x_1 + b_2 x_2 + c_2 x_3 + \cdots + f_2 x_i + \cdots + n_2 x_n + l_2\}, \text{etc. for}$$

all the m equations.

If we add all these together, we obtain

$$\frac{\partial(\Sigma v^2)}{\partial x_1} = (\Sigma a_j^2) x_1 + (\Sigma a_j b_j) x_2 + \cdots + (\Sigma a_j f_j) x_i + \cdots + (\Sigma a_j n_j) x_n + (\Sigma a_j l_j)$$

$$= 0 \quad \text{for minimum variation and therefore maximum probability.}$$

($\Sigma a_j f_j$ means " add all the values $a_1 f_1 + a_2 f_2 + \cdots + a_m f_m$ "). Similarly

$$\frac{\partial(\Sigma v_j^2)}{\partial x_2} = (\Sigma a_j b_j) x_1 + (\Sigma b_j^2) x_2 + \cdots + (\Sigma b_j f_j) x_i + \cdots + (\Sigma b_j n_j) x_n + (\Sigma b_j l_j) = 0$$

This gives us 'n' equations for the 'n' unknown adjusted values of x. If we write (Σa_j^2) as [aa], ($\Sigma a_j b_j$) as [ab], etc., for the sake of simplicity, then we can express these as in Table 11.3.

Thus,

$$[aa] = a_1^2 + a_2^2 + a_3^2 + \cdots + a_n^2, \quad [ab] = a_1 b_1 + a_2 b_2 + a_3 b_3 + \cdots + a_n b_n, \text{ etc.}$$

The equations in Table 11.3 are known as *normal equations*. Note that in these 'n' equations, the coefficients for 'x' are symmetrical about the diagonal, which itself is composed of the sum of the squares of the relevant coefficients.

Table 11.3 The Normal Equations

$$[aa]x_1 + [ab]x_2 + [ac]x_3 + \cdots + [af]x_i + \cdots + [an]x_n + [al] = 0$$
$$[ab]x_1 + [bb]x_2 + [bc]x_3 + \cdots + [bf]x_i + \cdots + [bn]x_n + [bl] = 0$$
$$[ac]x_1 + [bc]x_2 + [cc]x_3 + \cdots + [cf]x_i + \cdots + [cn]x_n + [cl] = 0$$
$$\vdots$$
$$[an]x_1 + [bn]x_2 + [cn]x_3 + \cdots + [fn]x_i + \cdots + [nn]x_n + [nl] = 0$$

We can write the condition equations in Table 11.2 as $\mathbf{MX + L = 0}$

$$\text{where } \mathbf{X} = \begin{pmatrix} x_1 \\ x_2 \\ x_3 \\ \cdots \\ x_n \end{pmatrix}; \quad \mathbf{L} = \begin{pmatrix} l_1 \\ l_2 \\ l_3 \\ \cdots \\ l_n \end{pmatrix} \text{ and } \mathbf{M} = \begin{pmatrix} a_1 & b_1 & c_1 & \cdots & n_1 \\ a_2 & b_2 & c_2 & \cdots & n_2 \\ a_3 & b_3 & c_3 & \cdots & n_3 \\ \cdots & \cdots & \cdots & \cdots & \cdots \\ a_m & b_m & c_m & \cdots & n_m \end{pmatrix}$$

Note that $\mathbf{M^T} = \begin{pmatrix} a_1 & b_1 & c_1 & \cdots & a_m \\ a_2 & b_2 & c_2 & \cdots & a_m \\ a_3 & b_3 & c_3 & \cdots & a_m \\ \cdots & \cdots & \cdots & \cdots & \cdots \\ n_m & n_m & n_m & \cdots & n_m \end{pmatrix}$ so that

$$\mathbf{M^T M} = \begin{pmatrix} [aa] & [ab] & [ac] & \cdots & [an] \\ [ab] & [bb] & [bc] & \cdots & [bn] \\ [ac] & [bc] & [cc] & \cdots & [cn] \\ \cdots & \cdots & \cdots & \cdots & \cdots \\ [an] & [bn] & [cn] & \cdots & [nn] \end{pmatrix}$$

Also $\mathbf{M^T L} = \begin{pmatrix} a_1 & b_1 & c_1 & \cdots & n_1 \\ a_2 & b_2 & c_2 & \cdots & n_2 \\ a_3 & b_3 & c_3 & \cdots & n_3 \\ \cdots & \cdots & \cdots & \cdots & \cdots \\ a_m & b_m & c_m & \cdots & n_m \end{pmatrix} * \begin{pmatrix} l_1 \\ l_2 \\ l_3 \\ \cdots \\ l_n \end{pmatrix} = \begin{pmatrix} [al] \\ [bl] \\ [cl] \\ \cdots \\ [nl] \end{pmatrix}$

Thus Table 11.3 can be written as $\mathbf{M^T(MX+L) = 0}$. Once we have formulated the necessary conditions it becomes a matter of routine to form and transform the relevant matrices and solve for the 'm' unknowns. If the 'm' equations are of unequal weight then instead of minimising Σv_j^2 we need to minimise $\Sigma w_j v_j^2$ and the normal equations become as shown in Table 11.4.

The above assumes that we have more equations ('m') than observed quantities ('n'), that is, $m > n$. It also assumes that each equation can be given an appropriate weight.

BEST-FIT SOLUTIONS

Table 11.4 Weighted Normal Equations

$$[waa]x_1 + [wab]x_2 + \cdots + [waf]x_i + \cdots + [wan]x_n + [wal] = 0$$
$$[wab]x_1 + [wbb]x_2 + \cdots + [wbf]x_i + \cdots + [wbn]x_n + [wbl] = 0$$
$$[wac]x_1 + [wbc]x_2 + \cdots + [wcf]x_i + \cdots + [wcn]x_n + [wcl] = 0$$

$$[wan]x_1 + [wbn]x_2 + \cdots + [wfn]x_i + \cdots + [wnn]x_n + [wnl] = 0$$

Box 11.8 — Solving for more Equations than Unknowns

Let

$$x + y + z - 9.2 = 0$$
$$x + y - z - 0.9 = 0$$
$$2x + 3y + 2z - 20.6 = 0$$
$$3x - 4y + 5z - 14.1 = 0$$
$$x + 5y - 3z - 4.9 = 0$$

We have 3 values (x, y, z) with 5 equations, assumed to be of equal weight. In matrix form,

$$\mathbf{A} = \begin{pmatrix} 1 & 1 & 1 \\ 1 & 1 & -1 \\ 2 & 3 & 2 \\ 3 & -4 & 5 \\ 1 & 5 & -3 \end{pmatrix}$$

Note $\mathbf{A}^T\mathbf{A} = \begin{pmatrix} 1 & 1 & 2 & 3 & 1 \\ 1 & 1 & 3 & -4 & 5 \\ 1 & -1 & 2 & 5 & -3 \end{pmatrix} * \begin{pmatrix} 1 & 1 & 1 \\ 1 & 1 & -1 \\ 2 & 3 & 2 \\ 3 & -4 & 5 \\ 1 & 5 & -3 \end{pmatrix}$

$$= \begin{pmatrix} 16 & 1 & 16 \\ 1 & 52 & -29 \\ 16 & -29 & 40 \end{pmatrix}$$

$$\mathbf{A}^T\mathbf{L} = \begin{pmatrix} 1 & 1 & 2 & 3 & 1 \\ 1 & 1 & 3 & -4 & 5 \\ 1 & -1 & 2 & 5 & -3 \end{pmatrix} * \begin{pmatrix} -9.2 \\ -0.9 \\ -20.6 \\ -14.1 \\ -4.9 \end{pmatrix} = \begin{pmatrix} -98.5 \\ -40 \\ -105.3 \end{pmatrix}$$

Thus

$$16x + y + 16z - 98.5 = 0$$
$$x + 52y - 29z - 40 = 0$$
$$16x - 29y + 40z - 105.3 = 0$$

From which the "best" values are:

$$x = 2.02, y = 2.93, z = 3.95$$

In much of geomatics, especially in position fixing, we deliberately make redundant observations in order to improve the accuracy of our work, both by identifying and thereby being able to eliminate any gross errors in measurement, but more particularly, by reducing the overall standard deviations in our measurements. This increases the probability of obtaining a better answer.

Consider then the case where we have more observations than there are conditions to satisfy. We must satisfy the basic conditions, while also ensuring that $\Sigma w_i \varepsilon_i^2$ is a minimum where 'w_i' is the weight of each observation and 'ε_i' is the correction to the observed quantity ('o_i'). 'i' is an observation that takes the value from one to 'n', where 'n' is the total number of observations.

If the best answer for each observation is 'x_i' then $x_i = o_i + \varepsilon_i$. If $\Sigma w_i \varepsilon_i^2$ is a minimum, then the differential with respect to 'ε' must be zero. Thus, $2w_1\varepsilon_1\delta_1\varepsilon_1 + 2w_2\varepsilon_2\delta\varepsilon_2 + \cdots + 2w_n\varepsilon_n\delta\varepsilon_n = 0$ or

$$w_1\varepsilon_1\delta\varepsilon_1 + w_2\varepsilon_2\delta\varepsilon_2 + \cdots + w_n\varepsilon_n\delta\varepsilon_n = 0$$

In the general case, we have 'n' observations of x and 'm' conditions to satisfy a set of equations as shown in Table 11.5. Here $x_i = o_i + \varepsilon_i$, where '$x$' is the adjusted value and 'o' is the observed value, the difference between them being the residual 'ε.'

Table 11.5 Observations and Conditions

$a_1(o_1 + \varepsilon_1)$	$+ b_1(o_2 + \varepsilon_2)$	$+ \cdots + f_1(o_i + \varepsilon_i)$	$+ \cdots + n_1(o_n + \varepsilon_n)$	$+ l_1$	$= 0$
$a_2(o_1 + \varepsilon_1)$	$+ b_2(o_2 + \varepsilon_2)$	$+ \cdots + f_2(o_i + \varepsilon_i)$	$+ \cdots + n_2(o_n + \varepsilon_n)$	$+ l_2$	$= 0$
$\vdots$					
$a_i(o_1 + \varepsilon_1)$	$+ b_i(o_2 + \varepsilon_2)$	$+ \cdots + f_i(o_i + \varepsilon_i)$	$+ \cdots + n_i(o_n + \varepsilon_n)$	$+ l_i$	$= 0$
$\vdots$					
$a_m(o_1 + \varepsilon_1)$	$+ b_m(o_2 + \varepsilon_2)$	$+ \cdots + f_m(o_i + \varepsilon_i)$	$+ \cdots + n_m(o_n + \varepsilon_n)$	$+ l_m$	$= 0$

The conditions must be fulfilled as best as possible, which means that if we put the observed quantities into the equations in Table 11.5, there will be a need for the 'n' unknowns to satisfy the following 'm' relationships; the quantities 'r' are the numerical values left over after trying to match the observed quantities with the conditions. These are shown in Table 11.6 where

$$a_i o_1 + b_i o_2 + c_i o_3 + \cdots + f_i o_i + \cdots + n_i o_n + l_i = r_i$$

Table 11.6 Relationships to be Optimized

$a_1\varepsilon_1$	$+ b_1\varepsilon_2$	$+ c_1\varepsilon_3$	$+ \cdots + f_1\varepsilon_i$	$+ \cdots + n_1\varepsilon_n$	$+ r_1$	$= 0$	(1)
$a_2\varepsilon_1$	$+ b_2\varepsilon_2$	$+ c_2\varepsilon_3$	$+ \cdots + f_2\varepsilon_i$	$+ \cdots + n_2\varepsilon_n$	$+ r_2$	$= 0$	(2)
$a_i\varepsilon_1$	$+ b_i\varepsilon_2$	$+ c_i\varepsilon_3$	$+ \cdots + f_i\varepsilon_i$	$+ \cdots + n_i\varepsilon_n$	$+ r_i$	$= 0$	(i)
$a_m\varepsilon_1$	$+ b_m\varepsilon_2$	$+ c_m\varepsilon_3$	$+ \cdots + f_m\varepsilon_i$	$+ \cdots + n_m\varepsilon_n$	$+ r_m$	$= 0$	(m)

BEST-FIT SOLUTIONS

Table 11.7 The Differentiated Equations

1.	$a_1 \delta\varepsilon_1$	$+ b_1 \delta\varepsilon_2$	$+ c_1 \delta\varepsilon_3$	$+ \cdots + f_1 \delta\varepsilon_i$	$+ \cdots + n_1 \delta\varepsilon_n$	$= 0$
2.	$a_2 \delta\varepsilon_1$	$+ b_2 \delta\varepsilon_2$	$+ c_2 \delta\varepsilon_3$	$+ \cdots + f_2 \delta\varepsilon_i$	$+ \cdots + n_2 \delta\varepsilon_n$	$= 0$
$\vdots$						
i.	$a_i \delta\varepsilon_1$	$+ b_i \delta\varepsilon_2$	$+ c_i \delta\varepsilon_3$	$+ \cdots + f_i \delta\varepsilon_i$	$+ \cdots + n_i \delta\varepsilon_n$	$= 0$
m.	$a_m \delta\varepsilon_1$	$+ b_m \delta\varepsilon_2$	$+ c_m \delta\varepsilon_3$	$+ \cdots + f_m \delta\varepsilon_i$	$+ \cdots + n_m \delta\varepsilon_n$	$= 0$

If we differentiate the equations in Table 11.6, assuming that ε is the variable, then we derive the equations shown in Table 11.7.

Next we introduce 'm' constants $K_1, K_2, \ldots, K_m$ and apply each of these to its equivalent differentiated equation in Table 11.7 and combine all the equations together by adding them. Thus, the equations become $K1*(\text{Equation (1)}) + K2*(\text{Equation (2)}) + \cdots$, etc., giving us

$$K_1(a_1 \delta\varepsilon_1 + b_1 \delta\varepsilon_2 + c_1 \delta\varepsilon_3 + \cdots + f_1 \delta\varepsilon_i + \cdots + n_1 \delta\varepsilon_n)$$
$$+ K_2(a_2 \delta\varepsilon_1 + b_2 \delta\varepsilon_2 + c_2 \delta\varepsilon_3 + \cdots + f_2 \delta\varepsilon_i + \cdots + n_2 \delta\varepsilon_n)$$
$$\vdots$$
$$+ K_m(a_m \delta\varepsilon_1 + b_m \delta\varepsilon_2 + c_m \delta\varepsilon_3 + \cdots + f_m \delta\varepsilon_i + \cdots + n_m \delta\varepsilon_n) = 0$$

Hence on rearranging terms

$$(a_1 K_1 + a_2 K_2 + \cdots + a_m K_m) \delta\varepsilon_1$$
$$+ (b_1 K_1 + b_2 K_2 + \cdots + b_m K_m) \delta\varepsilon_2$$
$$\vdots$$
$$+ (n_1 K_1 + n_2 K_2 + \cdots + n_m K_m) \delta\varepsilon_n = 0$$

For this to be identical to the condition that the weighted sum of the squares of the residuals is a minimum, namely:

$$\omega_1 \varepsilon_1 \delta\varepsilon_1 + \omega_2 \varepsilon_2 \delta\varepsilon_2 + \cdots + \omega_n \varepsilon_n \delta\varepsilon_n + = 0$$

we must have the relationships shown in Table 11.8.

Table 11.8 The Relationships between the Correlatives

$(a_1 K_1 + a_2 K_2 + \cdots + a_m K_m)$	$= w_1 \varepsilon_1$
$(b_1 K_1 + b_2 K_2 + \cdots + b_m K_m)$	$= w_2 \varepsilon_2$
$(n_1 K_1 + n_2 K_2 + \cdots + n_m K_m)$	$= w_n \varepsilon_n$

This gives us 'm' equations linking the 'm' unknown quantities 'K' with the unknown quanties 'ε.' These quantities K are called the *correlatives*. Substituting the values of 'ε' into our original equations, we obtain the equations for the correlatives given in Table 11.9, which can be rearranged as shown in Table 11.10.

Table 11.9   The Equations for the Correlatives

1. $a_1(a_1K_1 + a_2K_2 + \cdots + a_mK_m)/w_1$
   $+ b_1(b_1K_1 + b_2K_2 + \cdots + b_mK_m)/w_2 + \cdots + r_1 = 0$
2. $a_2(a_1K_1 + a_2K_2 + \cdots + a_mK_m)/w_1$
   $+ b_2(b_1K_1 + b_2K_2 + \cdots + b_mK_m)/w_2 + \cdots + r_2 = 0$

m. $a_m(a_1K_1 + a_2K_2 + \cdots + a_mK_m)/w_1$
   $+ b_m(b_1K_1 + b_2K_2 + \cdots + b_mK_m)/w_2 + \cdots + r_m = 0$

Table 11.10   Solving for the Correlatives

$$
\begin{aligned}
1. \quad & K_1(a_1a_1/w_1) + b_1b_1/w_2 + \cdots + n_1n_1/w_n) \\
+ \quad & K_2(a_1a_2/w_1) + b_1b_2/w_2 + \cdots + n_1n_2/w_n) \\
\vdots \\
+ \quad & K_m(a_1a_m/w_1) + b_1b_m/w_2 + \cdots + n_1n_m/w_n) + r_1 = 0 \\
2. \quad & K_1(a_1a_2/w_1) + b_1b_2/w_2 + \cdots + n_1n_2/w_n) \\
+ \quad & K_2(a_2a_2/w_1) + b_2b_2/w_2 + \cdots + n_2n_2/w_n) \\
\vdots \\
+ \quad & K_m(a_2a_m/w_1) + b_2b_m/w_2 + \cdots + n_2n_m/w_n) + r_2 = 0 \\
m. \quad & K_1(a_1a_m/w_1) + b_1b_m/w_2 + \cdots + n_1n_m/w_n) \\
+ \quad & K_2(a_2a_m/w_1) + b_2b_m/w_2 + \cdots + n_2n_m/w_n) \\
\vdots \\
+ \quad & K_m(a_ma_m/w_1) + b_mb_m/w_2 + \cdots + n_mn_m/w_n) + r_m = 0
\end{aligned}
$$

Since all this is very abstract, let us consider an example. If all the eight internal angles are measured in Figure 11.4 (a figure known to surveyors as a braced quadrilateral), then there will be redundant measurements. In fact, if A and B are known, then only two measures are required to fix C and two to fix D; hence, if we have eight measurements, there are four extra observations. These can be used to improve the accuracy of the determinations of C and D from A and B. If we take the braced quadrilateral in Figure 11.4(a), then obviously

$$\text{Angles } 1 + 2 + 3 + 8 = 180$$

$$\text{Angles } 2 + 3 + 4 + 5 = 180$$

$$\text{Angles } 4 + 5 + 6 + 7 = 180$$

$$\text{Angles } 6 + 7 + 8 + 1 = 180$$

$$\text{Angles } 1 + 2 + 3 + 4 + 5 + 6 + 7 + 8 = 360$$

We have five equations and four redundant observations, since if A and B are known, then angles 1, 2, 7, and 8 are sufficient to fix C and D; angles 3, 4, 5, and 6 are redundant. However, these five equations are not independent, since the fifth equation is the sum of the first and second, and also of the third and fourth. Furthermore, $\angle 4 + \angle 5 = \angle 1 + \angle 8$ while $\angle 2 + \angle 3 = \angle 6 + \angle 7$ and we cannot distinguish between the pairs. The angles in Figure 11.4(b) are the same as in Figure 11.4(a); but the figure is geometrically different.

BEST-FIT SOLUTIONS

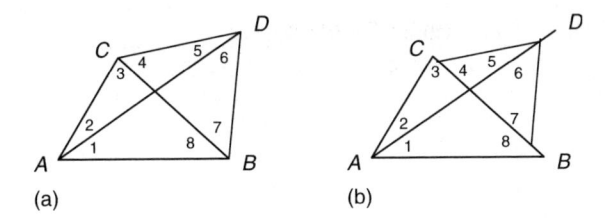

Figure 11.4 The braced quadrilateral.

The conclusion from all this is that care must be taken in defining the conditions that must be satisfied. In the case of the braced quadrilateral, in order to satisfy the geometry, we need to make sure that the scale is consistent. Since

$$AB/AC = \sin 3/\sin 8; \quad AC/CD = \sin 5/\sin 2$$

$$CD/DB = \sin 7/\sin 4 \text{ and } DB/AB = \sin 1/\sin 6$$

Then on multiplying all these together, we obtain

$$\frac{AB}{AC} * \frac{AC}{CD} * \frac{CD}{DB} * \frac{DB}{AB} = 1$$

Hence

$$\frac{\sin 3 * \sin 5 * \sin 7 * \sin 1}{\sin 8 * \sin 2 * \sin 4 * \sin 6} = 1$$

or

$$\sin 3 * \sin 5 * \sin 7 * \sin 1 = \sin 8 * \sin 2 * \sin 4 * \sin 6$$

So now we have

$$(4 + 5) = (1 + 8)$$

$$(2 + 3) = (6 + 7) \text{ ; and}$$

$$(1 + 2 + 3 + 4 + 5 + 6 + 7 + 8) = 360$$

These, together with the relationship between the sines of the angles, give four equations that must be satisfied in order that the geometry is robust. These equations are the condition equations and must be satisfied exactly.

The first three equations are of linear form, but at present the fourth is not. Let each observed angle have a correction ε_i seconds, $i = 1, \ldots, 8$. Then, Box 11.9 establishes a relationship based on the log sine conditions by reducing these to a set of linear relationships. Box 11.10 then summarizes the corrections that must be optimized.

Box 11.9 — Angles in a Braced Quadrilateral

Angle	Observed	log sine *	Δ1"	log sine*	Δ1"
1.	52° 48' 12.3"	−0.0987783	16		
2.	89 33 15.8			−0.0000131	0
3.	15 13 58.9	−0.5804649	77		
4.	17 23 14.0			−0.5245787	67
5.	57 49 31.7	−0.0724090	13		
6.	68 09 09.5			−0.0323685	8
7.	36 38 19.6	−0.2241944	28		
8.	22 24 31.0			−0.4188365	51
Σ = 360° 00' 12.8"		−0.9758466		−0.9757968	

Box 11.10 — An example of a Survey Adjustment

Using the data in Box 11.9:
  angles 1 + 8 = 75°12' 43.3" and 4 + 5 = 75° 12' 45.7"
Hence $\varepsilon_1 + \varepsilon_8 + 75° 12' 43.3" = \varepsilon_4 + \varepsilon_5 + 75° 12' 45.7"$

$$\varepsilon_1 - \varepsilon_4 - \varepsilon_5 + \varepsilon_8 - 2.4 = 0 \qquad (1)$$

Also angles 2 + 3 = 104° 47' 14.7" and 6 + 7 = 104° 47' 29.1"

$$\varepsilon_2 + \varepsilon_3 - \varepsilon_6 - \varepsilon_7 - 14.4 = 0 \qquad (2)$$

$$\varepsilon_1 + \varepsilon_2 + \varepsilon_3 + \varepsilon_4 + \varepsilon_5 + \varepsilon_6 + \varepsilon_7 + \varepsilon_8 + 12.8 = 0 \qquad (3)$$

Also

$$\sin 1 * \sin 3 * \sin 5 * \sin 7 = \sin 2 * \sin 4 * \sin 6 * \sin 8$$

or

$$\text{Log sin } 1 + \log \sin 3 + \log \sin 5 + \log \sin 7$$
$$= \log \sin 2 + \log \sin 4 + \log \sin 6 + \log \sin 8$$

The log (sin) values of the four odd numbered angles add up to −0.9758466 and the evens add up to −0.9757968. The change in the log sine that arises from a change of 1" can be obtained from tables or compared using a pocket calculator. (e.g., log sin (52° 48' 12.3") = −0.0987783 while log sin (52° 48' 13.3") = −0.0987767, which differs by $16*10^{-7}$). Thus from Box 11.9.

—Continued

BEST-FIT SOLUTIONS 193

> **Box 11.10 — An example of a Survey Adjustment (Continued)**
>
> $$16\varepsilon_1 + 77\varepsilon_3 + 13\varepsilon_5 + 28\varepsilon_7 - 9758466$$
>
> $$= 67\varepsilon_4 + 8\varepsilon_6 + 51\varepsilon_8 - 9757968$$
>
> or
>
> $$16\varepsilon_1 + 77\varepsilon_3 - 67\varepsilon_4 + 13\varepsilon_5 - 8\varepsilon_6 + 28\varepsilon_7 - 51\varepsilon_8 - 498 = 0$$
>
> Hence, we have
>
> $$16\varepsilon_1 + 77\varepsilon_3 - 67\varepsilon_4 + 13\varepsilon_5 - 8\varepsilon_6 + 28\varepsilon_7 - 51\varepsilon_8 - 498 = 0 \quad (4)$$
>
> This gives four linear equations (1) to (4)

In Box 11.9, the logarithm of the sine of each of the eight angles has been tabulated together with the differences in each logarithm for one second of arc $\Delta 1''$. This allows us to express the errors in terms of linear combinations derived from a relationship that was expressed as a series of sines that were multiplied together.

Box 11.10 shows how the process is carried out, leading to a simple linear relationship between the corrections to be applied to each of the angles. The box also shows the simple relationships between the sums of the angles. There are four independent equations relating to 8 unknowns. In addition, we want $\Sigma\varepsilon^2$ to be a minimum. Using the formulae derived above, $n = 8$ and $m = 4$, we can tabulate the four equations $E1$, $E2$, $E3$, and $E4$ and 8 unknowns as in Table 11.11.

Table 11.11   The Correlative Matrix

	$a=\varepsilon_1$	$b=\varepsilon_2$	$c=\varepsilon_3$	$d=\varepsilon_4$	$e=\varepsilon_5$	$f=\varepsilon_6$	$g=\varepsilon_7$	$h=\varepsilon_8$	r
E1	1	0	0	−1	−1	0	0	1	−2.4
E2	0	1	1	0	0	−1	−1	0	−14.4
E3	1	1	1	1	1	1	1	1	12.8
E4	16	0	77	−67	13	−8	28	−51	−498

In Table 11.11, the rows are the condition equations and the columns the corrections. Thus, for example, the row $E1$ shows that

$$1\varepsilon_1 - 1*\varepsilon_4 - 1*\varepsilon_5 + 1*\varepsilon_8 - 2.4 = 0$$

Using the equations in Table 11.10 and assuming the weights all equal 1, we can substitute numbers for all the expressions such as

$$K_1(a_1a_1/w_1 + \cdots + n_1n_1/w_n) + K_2(a_1a_2/w_1 + \cdots + n_1n_2/w_n), \text{ etc.}$$

Thus, in the first equation, the coefficient of

K_1 from $(E1)^2$ is $1^2 + 0^2 + 0^2 + (-1)^2 + (-1)^2 + 0^2 + 0^2 + 1^2 = 4$

K_2 from $(E1*E2)$ is $1*0 + 0*1 + 0*1 - 1*0 - 1*0 + 0*(-1)$
$+ 0*(-1) + 1*0 = 0$

K_3 comes from $E1*E3$ and is $1 + 0 + 0 - 1 - 1 + 0 + 0 + 1 = 0$

K_4 from $E1*E4$ is $16 + 0 + 0 + 67 - 13 + 0 + 0 - 51 = 19$

Hence, using the numbers in Table 11.11 and substituting them in to the equations given in Table 11.10, we obtain the following:

1. $4*K_1 + 0*K_2 + 0*K_3 + 19*K_4 - 2.4 = 0$

2. $0*K_1 + 4*K_2 + 0*K_3 + 57*K_4 - 14.4 = 0$

3. $0*K_1 + 0*K_2 + 8*K_3 + 8*K_4 + 12.8 = 0$

4. $19*K_1 + 57*K_2 + 8*K_3 + 14292*K_4 - 498 = 0$

From 1, $K_1 = 0.6 - 4.75 K_4$. From 2, $K_2 = 3.6 - 14.25 K_4$.
From 3, $K_3 = -1.6 - K_4$. From 4, $K_4 = (294.2)/(13381.5) = 0.022$. So

$$K_1 = 0.496; K_2 = 3.287; K_3 = -1.622; K_4 = 0.022$$

Substituting back, using Table 11.11 and reading the columns downwards so that under the column headed "$a = \varepsilon_1$," we have

$$\varepsilon_1 = 1*E1 + 0*E2 + 1*E3 + 16*E4,$$

$$\varepsilon_2 = 0*E1 + 1*E2 + 1*E3 + 0*E4,$$

$$\varepsilon_3 = 0*E1 + 1*E2 + 1*E3 + 77*E4, \cdots$$

The overall results are given in Box 11.11. The somewhat tedious tortuous calculations have led to a solution in which $\Sigma\varepsilon^2 = 79.96$. This is the minimum least-square value. Many types of observation can be processed in this way, provided we know the conditions that must be fulfilled and the relative weighting of the different observations. Although tedious to go through by long-hand, once the equations are obtained the electronic computer can take over and provide a most probable value for each of the observations.

BEST-FIT SOLUTIONS

> **Box 11.11 — The Final Solution**
>
> $$\varepsilon_1 = 1*K_1 + 0*K_2 + 1*K_3 + 16*K_4 = -0.774 = -0.8$$
>
> Angle 1 = 52 48 11.5
>
> $$\varepsilon_2 = 0*K_1 + 1*K_2 + 1*K_3 + 0*K_4 = 1.665 = +1.7$$
>
> Angle 2 = 89 33 17.5
>
> $$\varepsilon_3 = 0*K_1 + 1*K_2 + 1*K_3 + 77*K_4 = 3.359 = 3.3$$
>
> Angle 3 = 15 14 02.2
>
> $$\varepsilon_4 = -1*K_1 + 0*K_2 + 1*K_3 - 67*K_4 = -3.593 = -3.6$$
>
> Angle 4 = 17 23 10.4
>
> $$\varepsilon_5 = -1*K_1 + 0*K_2 + 1*K_3 + 13*K_4 = -1.832 = -1.8$$
>
> Angle 5 = 57 49 29.9
>
> $$\varepsilon_6 = 0*K_1 + -1*K_2 + 1*K_3 - 8*K_4 = -5.085 = -5.1$$
>
> Angle 6 = 68 09 04.4
>
> $$\varepsilon_7 = 0*K_1 + -1*K_2 + 1*K_3 + 28*K_4 = -4.293 = -4.3$$
>
> Angle 7 = 36 38 15.3
>
> $$\varepsilon_8 = 1*K_1 + 0*K_2 + 1*K_3 - 51*K_4 = -2.248 = -2.2$$
>
> Angle 8 = 22 24 28.8

The technique of least-square adjustments has many applications in surveying and mapping, and indeed other areas where a best-fit solution is required. The algorithms are ideally suited to electronic data processing; but the quality of the results will depend on the quality of the observations and the assumptions underlying the adjustment procedures. If, for example, each observation is not truly independent and is somehow correlated with other observations, then the underlying assumptions of the least-square process are not fully valid. Techniques are available to test for this using variance and covariance analysis but are beyond the scope of the present text.

Further Reading

There are many excellent books covering basic mathematics and more specific techniques in topics such as algebra, calculus, geometry and statistical analysis. The following is a small selection of some that are available.

GIS

Longley, P.A., Goodchild, M.F., Maguire, D.J. & Rhind, D.W. *2001. Geographical Information: Systems and Science*, New York: Wiley

Worboys, M. *2004. Geographic Information Systems: A Computing Perspective, 3rd edition,* London: Taylor and Francis

General Mathematics

Allan, A.L. *2004. Maths for Map Makers,* 2nd edition, Whittles Publishing

Borowski, E.J. & Borwein, J.M. *2002. Dictionary of Mathematics*, 2nd Edition, HarperCollins

Cox, W. *2001. Understanding Engineering Mathematics*, Oxford: Butterworth Heineman

Specific Topics

Angell, I.O. *1981. A Practical Introduction to Computer Graphics*, New York: Macmillan

Iliffe, J.C. *2002. Datums and Map Projections*, Whittles Publishing

Jackson, J.E. *1987. Sphere, Spheroid and Projections for Surveyors*, Oxford: Blackwell Scientific Publications

Professional Books

Lancaster, P. & Salkausus, K. *1986. Curve and Surface Fitting—An Introduction*, New York: Academic Press

Mikhail, E., Bethel, J. & McGlone, J.C. *2001. Introduction to Modern Photogrammetry*, New York: John Wiley and Sons

Thompson, E.H. *1969. An Introduction to the Algebra of Matrices with Some Applications*, Bristol: Adam Hilger

Wolf, P.R & Ghilani, G.D. *1997. Adjustment Computations—Statistics and Least Squares in Surveying and GIS*, New York: John Wiley & Sons

Index

A

Accuracy 6
Algebra (chapter 3)
Analysis of variance (ANOVA) 166–169
Angles 5
  acute 44
  obtuse 44, 63–65
  on a sphere 72–76
  subtended by an arc 45
Areas
  by integration 87–89
  of irregular shapes 88, 89
  of polygons 48, 49
  of triangles 46–48
Argument 36
Arithmetic (chapter 2)
  axioms of 10
  rules of 9–13
Arrays 91

B

Bearings 5
  and distances 67–71
Bezier curves 128–130
Binary system 13–14
Binomial expansion 154–156, 159, 160

C

Calculus (chapter 6)
  differential 77
  integral 77
Cassini 144
Central limit theorem 161
Central tendency 156–160
Centroid 39, 49

Chi-squared test 169–170
Chords 45
Circles
  auxiliary 54, 117
  equation of 33
  great and small 53–54
Circumcenter 50
Clipping 28
Co-latitude 75–76, 146
Condition equations 188
Conformal projections 144
Conic sections 55–57
Convergence 36
Co-ordinates
  from observed angles 69
  homogeneous 131
  polar 5
  rectangular Cartesian 4–5
Correlation 173–175
  coefficient of 174
Correlatives 189
Cosine formula
  on a plane surface 62
  on a sphere 74
Covariance 174
Cubic functions 36, 80, 122, 124–126
Curvature - radius of 121–122
Curves and Surfaces (chapter 8)
  Bezier 128–130
  curve fitting 122–130
  parametric forms 115–118

D

Data
  categorical 2
  definition of 1
  integer 3

Data (*continued*)
  interval 3
  nominal 2
  numerical 2
  ordinal 2
Degrees of freedom 164
Delaunay 53
Denominator 12
Determinants 94–95, 98
Differentiation 77–85
  derivatives (2nd, 3rd, 4th) 79
  partial 85, 184
  of trigonometric functions 81–83
Direction cosines 110

E

Element in a matrix 91
Elemental triangle 141, 143, 146, 148
Ellipse 54–55, 116–120
  axes (major, minor) 54
  auxiliary circle 54
  directrix 55
  eccentricity 55
  focus 55
  radius (nu) 119, 120
Ellipsoid 6, 54
Equal area projection 143
Equations
  conditional 32
  identical 32
  quadratic 33–35
  simultaneous 33, 35
Errors Type I & II 163–164
Error term 177
Euclid 43
Euler number ('e') 18, 84
Extrapolation 38, 39

F

Factors 12, 25
F–test 166, 167
Functions 36–38
  inverse 37
  polynomial 83–85

G

Geographic information (chapter 1)
Geographic Information Systems 1
Geometry (chapter 4)
Gosset ('student') 165
Gradients 115
Graphs 37–38

H

Heuristic 154
Hidden lines and surfaces 139–140
Histograms 156
Homogeneous co-ordinates 131
Hyperbola 56, 57

I

Incenter 50
Indices 17–21
Inflection point 80, 81
Information 1
Integration 85–90
  constant of 85
  definite 88
  indefinite 85
  of trigonometric functions 86–87
Interaction 169
Intercepts 25
Interpolation 38–41
Iteration 15–16

L

Latitude 6, 54
  geodetic 119
  reduced 119
Least square solution 177, 184–195
Left hand rule 103
Linearization 182–184
Lines
  equations of 25–29, 36
  intersecting 25–29
  orthogonal 116
  parallel 28, 132
Logarithms 18–21
Longitude 6, 54

M

Map projections 140–151
  azimuthal/zenithal 142, 145–147
  Cassini 144
  conformal/orthomorphic 144
  conical 142, 147–151
  cylindrical 142–145
  equal area 143
  Mercator's 144
  stereographic 146
  Transverse Mercator 144–145
Matrices (chapter 7)
  addition and subtraction 92, 93
  adjugate 97
  augmented 109

of coefficients 109
cofactor 96
diagonal 92
identity 92
inverse 94, 98
minor 96
multiplication 93, 94
orthogonal 101
partitioned 105
rotation 99–105
similarity 101
square 91
sub-matrix 96
transpose 95
triangular 106
Maxima and minima 80
McLaurin's theorem 84, 183
Mean value 157–159
  geometric 174
  weighted 180, 181
Measurement scale
  absolute 3
  interval 3
  nominal 2
  ordinal 2
  ratio 3
Median 157
Mercator 144
Mode 157
Modulus
  of any number 33
  of logarithms 20

N

Newton's method 182
Nodes 123
Non-parametric tests 169
Normal distribution 160–163
Normal equations 186
Normal to curve 51, 79
Null hypothesis 163
Numbers
  antilogarithm 21
  base of logarithms 18
  binary digits 13
  decimal 12–14
  factorials 18
  fractions 12
  highest common factor 12
  imaginary 12
  indices 17
  integer 3, 11
  logarithms 17–21
  lowest common denominator 12
  modulus of logarithms 20

  Natural/Naperian logarithms 18
  perfect squares 15, 34
  power of 17
  rational 12
  real 11
  rounding up and down 16
  significant figures 16
Numerator 12

O

Observation equations 188
Origins
  false 26
  translation of 99–100
  true 4, 25
Orthocenter 50
Orthogonal
  lines 116
  matrices 101
Orthomorphism 144

P

Parabola 56, 57
Parallelepiped 112
Parametric
  forms 115–118
  tests 169
Photogrammetry 104
Piecewise curves 123
Pixel 8
Planes
  equation of 31, 113–114
  intersection of 32
Points at infinity 132, 139
Point-in-polygon 30, 31
Poisson distribution 170–172
Polynomials 36, 83–85
Polygons 51–53
  areas of 48–49
  Theissen 41, 53
Populations 153
Precision 6
Probability 153–156
Pythagoras 15, 23, 24

Q

Quadratics 33, 34

R

Radius of curvature 121–122
Random numbers 154
Raster image 7, 8, 37

Regression 176–180
Residuals 157, 179
Rhumb lines 144
Right hand rule 103
Rotation of axes 99–105
  of objects 132–139

S

Samples 153
Scalar 109
Scale, *see* measurement scale
Scale factor 29
Segments of a circle 45
Series 36
Significance 163–164
  1, 2 tailed 167
Simpson's rule 88–89
Sine formula:
  plane surface 62
  spherical angles 73
Singular determinant 94
Spatial referencing 4–6
Spheres 53–54
Spherical excess 54
Spline 123
Squares 15
  difference of two 24
  perfect 15, 34
Square roots 15, 16
Standard
  deviation 157–160
  parallels 147–151
Statistics chap 10
  descriptive 153
  inferential 153
Stochastic models 154
Strings 126

T

't' Test 165–166
Tangent
  of angle 59–61
  to curve 51, 77, 115
Taylor's theorem 183
Theissen polygons 41, 53
Topology 8
  adjacency 8
  containment 8
  connectivity 8
Transformations (chapter 9)
  affine 137
  perspective 138, 140
  similarity 137
Translation 99–100
Trapezium 46

Traverses 69–72
Trend surface 179
Triangles 29, 43–51
  areas of 46–48
  congruent 44
  Delaunay 53
  equilateral 44
  isosceles 44
  scalene 44
  similar 29, 44
  spherical 72
Triangulation
  of height points 41
  networks 52
Trigonometric functions
  arc-sin, arc-cos, arc-tan 61
  combined angles 65–67
  cosec, cot, sec 61
  cosine formula - plane triangle 62
  cosine formula - spherical triangle 74
  cosine, sine, tangent 60–67
  obtuse angles 63–65
  sine formula - plane triangle 62
  sine formula - spherical triangle 73
Trigonometry (chapter 5)
Type I and Type II errors 163–164

U

Unbiased estimates 164

V

Vanishing points 132
Variables
  arguments for 36
  continuous 3
  dependent/independent 27
  discrete 3
Variance 157–160, 168, 180
  analysis of 166–169
Variation 168
Vectors 7, 109–114
  of constants 109
  cross product 111–113
  dot product 111–113
  scalar triple product 112–114
  unit 110
Volumes 90

W

Weights 180–181, 187
Windows (clipping) 28

Z

Z score 157